OECD-FAO Agricultural Outlook 2016-2025

Please cite this publication as:
OECD/FAO (2016), *OECD-FAO Agricultural Outlook 2016-2025*, OECD Publishing, Paris.
http://dx.doi.org/10.1787/agr_outlook-2016-en

ISBN 978-92-64-25322-3 (print)
ISBN 978-92-64-25323-0 (PDF)

OECD-FAO Agricultural Outlook
ISSN 1563-0447
ISSN 1999-1142

FAO:
ISBN 978-92-5-109296-5

Photos credits: Cover © Original cover concept designed by Juan Luis Salazar. Adaptations by OECD.

Corrigenda to OECD publications may be found on line at: *www.oecd.org/about/publishing/corrigenda.htm*.

Foreword

The food and agriculture systems fulfil a wide range of functions vital to the well-being of humanity. Being at the heart of global food security, they are expected to provide the world with adequate and reliable supplies of safe, healthy and nutritious food. They are also crucial for the livelihoods of billions of people, including many of the world's poorest, providing direct employment and income and contributing to the broader rural and overall economic development. To continue to fulfil these key roles, agricultural productivity has to increase in a sustainable manner.

The 2016 edition of the OECD-FAO Agricultural Outlook *provides an assessment of the medium-term prospects of global agriculture. The report highlights that for the sector to meet the expanding demand for food, feed and raw products for industrial uses, significant production growth is needed. This expansion will have to take place in the face of declining land and water availability for many areas in the world, compounded by the effects of climate change. It is clear that the majority of growth will have to come from more efficiency in agricultural production, but also from improvements throughout the wider value chain.*

The international community has recognised the key role of agriculture in addressing society's goals. Agriculture is a key sector for the achievement of many goals in the 2030 Agenda for Sustainable Development, which aims to end poverty and hunger and promote prosperity and people's wellbeing, while protecting the environment. This Outlook *outlines how agriculture can actively contribute to the attainment of these goals.*

While improvements to the global availability of, and access to, food are expected in the coming years, many countries will continue to be burdened with undernourishment and face increasingly complex issues of various forms of malnutrition, as was highlighted by the 2014 Second International Conference on Nutrition's Declaration. Moreover, stability and reliability of food supplies are also at risk in a diverse range of geographical regions and climate zones, due primarily to the effects of climate change. Recognising the vulnerability of food production systems to the adverse impacts of climate change, 195 countries agreed to take measures under the Paris Agreement reached at the 21st Conference of the Parties (COP21) in December 2015. Agriculture and the food chain will need to adapt to the changing climate and contribute to the mitigation efforts.

This Outlook *also confirms the increasing role of trade in global food security, as food supply and demand will be more and more geographically separated. Reliable trade relations between import dependent countries and their suppliers are of vital importance. The Nairobi Package, adopted at the 10th WTO Ministerial Conference in December 2015, contains decisions that constitute an important step in the reform of agricultural trade, in particular concerning export competition and issues such as the elimination of export subsidies and disciplines on export credits, food aid and state-trading enterprises.*

All these wide-ranging expectations for food and agriculture were at the heart of the discussions and the shared goals expressed by Ministers at the OECD Agriculture Ministerial held in April 2016, under the theme of "Better policies to achieve a productive, sustainable and resilient global food system".

This year's edition of the Outlook includes a special focus on the prospects and challenges for agriculture in Sub-Saharan Africa. The region is home for nearly 1 billion people, and agriculture remains a crucial sector for providing livelihoods to the majority of households. The report provides comprehensive projections for agricultural production and demand for more than 20 agricultural commodities particularly important to Sub-Saharan Africa. The Outlook is generally positive, yet the challenge of feeding rapidly rising populations remains formidable. The region has to overcome the challenge of low productivity of agricultural resources in the face of rapid urbanisation, increased globalization, the impacts of climate change, changing diets and the need for creating employment opportunities. The Outlook identifies some strategic priorities for ensuring that the region can take advantage of the opportunities and face the challenges ahead to achieve sustainable agri-food systems.

We believe that our collaborative effort on the annual production of the Agricultural Outlook, and also on the recently published OECD-FAO Guidance for Responsible Agricultural Supply Chains, enhance stakeholders' understanding of the complexity of agriculture and the food system in general. This work provides a plausible scenario of world agriculture over the coming decade, which can serve to inform and support efforts by governments and other actors in taking appropriate action to the benefit of our societies.

José Graziano da Silva,
Director-General
Food and Agriculture Organization
of the United Nations

Angel Gurría,
Secretary-General
Organisation for Economic
Co-operation and Development

Preface

The Agricultural Outlook, 2016-2025, is a collaborative effort of the Organisation for Economic Co-operation and Development (OECD) and the Food and Agriculture Organization (FAO) of the United Nations. It brings together the commodity, policy and country expertise of both organisations and input from collaborating member countries to provide an annual assessment of prospects for the coming decade of national, regional and global agricultural commodity markets. The special feature on Sub-Saharan Africa has been prepared in collaboration with analysts associated with the Regional Network of Agricultural Policy Research Institutes (ReNAPRI) and Bureau for Food and Agricultural Policy (BFAP). However, OECD and FAO are responsible for the information and projections contained in this document, and the views expressed in the special feature do not necessarily reflect those of ReNAPRI and BFAP.

The baseline projection is not a forecast about the future, but rather a plausible scenario based on specific assumptions regarding the macroeconomic conditions, the agriculture and trade policy settings, weather conditions, longer term productivity trends and international market developments. The projections of production, consumption, stocks, trade and prices for the different agricultural products described and analysed in this report cover the years 2016 to 2025. The evolution of markets over the outlook period is typically described using annual growth rates or percentage changes for the final year 2025 relative to a three-year base period of 2013-15.

The individual commodity projections are subject to critical examination by experts from national institutions in collaborating countries and international commodity organisations prior to their finalisation and publication in this report. The risks and uncertainties around the baseline projections are examined through a number of possible alternative scenarios and stochastic analysis, which illustrate how market outcomes may differ from the deterministic baseline projections.

The complete Agricultural Outlook, including more detailed commodity chapters, the full statistical annex and fully documented Outlook database, including historical data and projections, can be accessed through the OECD-FAO joint internet site: www.agri-outlook.org. The published Agricultural Outlook 2016-2025 report provides: an overview of global agriculture and prospects; an in-depth analysis of the outlook for Sub-Saharan Africa agriculture and a consideration of some of the challenges facing the sector; and two-page snapshots for each commodity with associated statistical tables. The more detailed commodity chapters are contained in the OECD's iLibrary version of the report.

Acknowledgements

The *Agricultural Outlook* is jointly prepared by the OECD and FAO Secretariats.

At the OECD, the baseline projections and *Outlook* report were prepared by members of the Agro-Food Trade and Markets Division of the Trade and Agriculture Directorate: Marcel Adenäuer, Jonathan Brooks (Head of Division), Annelies Deuss, Armelle Elasri (publication co-ordinator), Gen Furuhashi, Hubertus Gay (*Outlook* co-ordinator), Céline Giner, Gaëlle Gouarin, Pete Liapis, Claude Nenert, Graham Pilgrim and Grégoire Tallard. The OECD Secretariat is grateful for the contributions provided by visiting experts Selebogo Leshoro (South African Department of Agriculture, Forestry and Fisheries), John Saunders (Lincoln University, New Zealand), and Junye Zhao (Chinese Academy of Agricultural Sciences). The organisation of meetings and document preparation were provided by Helen Maguire and Özge Taneli-Ziemann. Technical assistance in the preparation of the *Outlook* database was provided by Eric Espinasse and Frano Ilicic. Many other colleagues in the OECD Secretariat and member country delegations provided useful comments on earlier drafts of the report.

At the Food and Agriculture Organization of the United Nations, the projections were prepared by economists and commodity officers from the Trade and Markets Division (EST) under the leadership of Boubaker Ben-Belhassen (EST Division Director) and with the overall guidance of Kostas Stamoulis (Assistant Director-General *ad interim,* Economic and Social Development Department). The core projections team consisted of ElMamoun Amrouk, Sergio René Araujo Enciso, Pedro Arias, Emily Carroll, Merritt Cluff, Hannah Fried, Yasmine Iqbal, Holger Matthey (Team Leader) and Jorge Soguero Escuer. Commodity expertise was provided by Abdolreza Abbassian, Concepcion Calpe, Kaison Chang, Michael Griffin, Shirley Mustafa, Adam Prakash, Shangnan Shui, and Peter Thoenes. We thank visiting expert Tracy Davids from the Bureau for Food and Agricultural Policy at the University of Pretoria. Stefania Vannuccini from the FAO Fisheries and Aquaculture Department contributed, with technical support from Pierre Charlebois. Advice on fishmeal and fish oil issues was provided by Enrico Bachis from the Marine Ingredients Organisation (IFFO). Research assistance and database preparation were provided by Claudio Cerquiglini, Julie Claro, Emanuele Marocco and Marco Milo. This edition also benefited from comments made by other colleagues from FAO and member country institutions and was closely reviewed by Günter Hemrich, Michelle Kendrick, Brave Ndisale, Coumba Sow, Kostas Stamoulis and Rob Vos. FAO's James Edge, Yongdong Fu, Pedro Javaloyes, Jessica Mathewson and Juan Luis Salazar provided invaluable assistance with publication and communication issues.

Chapter 2 of the *Outlook,* "Agriculture in Sub-Saharan Africa: Prospects and challenges for the next decade", was prepared by the technical teams at FAO and OECD, in collaboration with the Regional Network of Agricultural Policy Research Institutes (ReNAPRI). The contributions by ReNAPRI were led by Tracy Davids of the Bureau for Food and Agricultural Policy (BFAP), University of Pretoria, South Africa, with support from Sharolyn Arnett (ReNAPRI intern). Comments and support were provided by the following colleagues from FAO headquarters and decentralised offices (listed in alphabetical order): Carlo Angelico, Federica Angelucci, Jean Balié, Elisenda Estruch, Léopold Ghins, Dorian Kalamvrezos Navarro, Suffyan Koroma, Ekaterina Krivonos, Yanyun Li, Jonathan Pound, Paul Racionzer, Nancy Morgan, Jamie Morrison, Signe Nelgen, Jean Senahoun and Bukar Tijani. The box on "Emergence of a West African food economy" was prepared by Thomas Allen and Philipp Heinrigs of the Sahel and West Africa Club (SWAC).

Finally, information and feedback provided by the International Cotton Advisory Committee, International Dairy Federation, International Grains Council, International Sugar Organization and World Association of Beet and Cane Growers is gratefully acknowledged.

Table of Contents

Tables

Figures

Acronyms and abbreviations

ACP	African, Caribbean and Pacific countries
AD	Anti-dumping
AEC	ASEAN Economic Community
AfDB	African Development Bank
AMIS	Agricultural Market Information System
ARC	Agricultural Risk Coverage (US Farm Bill Instrument)
ASAP	Adaptation for Smallholder Agriculture Program
ASEAN	Association of South East Asian Nations
ASF	African Swine Fever
AU	African Union
BGI	Blue Growth Initiative
Bln	Billion
Bln L	Billion litres
BRIC	Emerging economies of Brazil, Russian Federation, India and China
BRICS	Emerging economies of Brazil, Russian Federation, India, China and South Africa
Bt	*Bacillius thuringiensis*
Bln t	Billion tonnes
CAADP	Comprehensive African Agricultural Development Programme
CAP	Common Agricultural Policy (European Union)
CCAFS	Climate Change, Agriculture and Food Security
CCC	Commodity Credit Corporation
CGIAR	Consultative Group of International Agricultural Research
ChAFTA	China-Australia Free Trade Agreement
CIF	Cost, insurance and freight
CIS	Commonwealth of Independent States
CO_2	Carbon dioxide
COMESA	Common Market for Eastern and Southern Africa
COP21	21st session of the Conference of the Parties; United Nations conference on climate change
CPI	Consumer Price Index
CPIF	Consumer Price Index for Food
CRP	Conservation Reserve Program (United States)
CSP	Conservation Stewardship Program (United States)
CTA	Technical Centre for Agricultural and Rural Cooperation
cts/lb	Cents per pound
CV	Coefficient of variation
CVD	Countervailing duty
c.w.e.	Carcass weight equivalent

DDA	Doha Development Agenda
DDGs	Dried Distiller's Grains
DNA	Deoxyribonucleic acid
DRC	Democratic Republic of the Congo
dw	Dressed weight
EAC	East African Community
EBA	Everything-But-Arms Initiative (European Union)
ECOWAS	Economic Community of West African States
EISA	Energy Independence and Security Act of 2007 (United States)
ELS	Extra Long Staple
EM-DAT	Emergency Events Database of the Centre for Research on the Epidemiology of Disasters
El Niño	Climatic condition associated with the temperature of major sea currents
EMEs	Emerging Market Economies
ENSO	El Niño Southern Oscillation
EPA	US Environmental Protection Agency
ERS	Economic Research Service of the US Department for Agriculture
EQUIP	Environmental Quality Incentives Program (United States)
ESA	Eastern and Southern Africa
est	Estimate
EU	European Union
EU15	Fifteen member states that joined the European Union before 2004
EU28	Twenty eight member states of the European Union
FAO	Food and Agriculture Organization of the United Nations
FDP	Fresh dairy products
FFV	Flex fuel Vehicles
FOB	Free on board (export price)
FMD	Foot and Mouth Disease
FTA	Free Trade Agreement
G-20	Group of 20 important developed and developing economies (see Glossary)
GDP	Gross domestic product
GDPD	Gross domestic product deflator
GHG	greenhouse gas
GIEWS	Global Information and Early Warning System on Food and Agriculture
GM	Genetically modified
ha	Hectares
HFCS	High fructose corn syrup
hl	Hectolitre
IARC	International Agency for Research on Cancer
IEA	International Energy Agency
IFA	International Fertilizer industry association
IFAD	International Fund for Agricultural Development
IFPRI	International Food Policy Research Institute
IGC	International Grains Council
ILUC	Indirect Land Use Change
IMF	International Monetary Fund
INDC	Intended Nationally Determined Contributions

IPCC	Intergovernmental Panel on Climate Change
IUU	Illegal, unreported and unregulated (fishing)
kg	Kilogrammes
kha	Thousand hectares
kt	Thousand tonnes
La Niña	Climatic condition associated with the temperature of major sea currents
LAC	Latin America and the Caribbean
lb	Pound
LDCs	Least Developed Countries
LPAA	Lima-Paris Action Agenda
lw	Live weight
MAFAP	Monitoring and Analysing Food and Agricultural Policies
MBM	Meat and bone meal
MDGs	Millennium Development Goals
MENA	Middle East and North Africa
MERCOSUR	Mercado Común del Sur / Common Market of South America
MFA	Multi-fibre Arrangement
Mha	Million hectares
mln	Million
Mn L	Million litres
MPS	Market Price Support
Mt	Million tonnes
N	Nitrogen
NAFTA	North American Free Trade Agreement
NEPAD	New Partnership for Africa's Development
OECD	Organisation for Economic Cooperation and Development
OIE	World Organisation for Animal Health
OLS	Ordinary Least Squares
OPEC	Organization of Petroleum Exporting Countries
P	Phosphorus
p.a.	Per annum
PCE	Private consumption expenditure
PEDv	Porcine Epidemic Diarrhoea virus
PLC	Price Loss Coverage (US Farm Bill instrument)
PoU	Prevalence of Undernourishment
PPI	Producer Price Index
PPP	Purchasing power parity
PSE	Producer Support Estimate
PSMA	Agreement on Port State Measures
RED	Renewable Energy Directive in the EU
RFS/RFS2	Renewable Fuels Standard in the US, which is part of the Energy Policy Act
RIN	Renewable Identification Numbers prices
rse	Raw sugar equivalent
RTA	Regional Trade Agreements
r.t.c.	Ready to cook
r.w.e.	Retail weight equivalent
SADC	Southern African Development Community

SDG	Sustainable Development Goals
SFP	Single Farm Payment (European Union)
SME	Small and Medium-sized Enterprises
SMP	Skim milk powder
SNAP	Supplemental Nutrition Assistance Program (United States)
SPS	Single payment scheme (European Union)
SSA	Sub-Saharan Africa
SSM	Special Safeguard Mechanism
SWAC	Sahel and West Africa Club
t	Tonnes
t/ha	Tonnes/hectare
TFP	Total Factor Productivity
TPP	Trans Pacific Partnership
tq	Tel quel basis
TRQ	Tariff rate quota
TTIP	Transatlantic Trade and Investment Partnership
UN	The United Nations
UNDP	United Nations Development Programme
UNFCCC	United Nations Framework Convention on Climate Change
URAA	Uruguay Round Agreement on Agriculture
US	United States
USDA	United States Department of Agriculture
WFP	World Food Programme
WHO	World Health Organization
WMP	Whole milk powder
wse	White sugar equivalent
WTO	World Trade Organization

Currencies

ARS	Argentinean peso	KRW	Korean won
AUD	Australian dollars	MXN	Mexican peso
BDT	Bangladeshi taka	MYR	Malaysian ringgit
BRL	Brazilian real	NZD	New Zealand dollar
CAD	Canadian dollar	PKR	Pakistani rupee
CLP	Chilean peso	RUB	Russian ruble
CNY	Chinese yuan renminbi	SAR	Saudi riyal
DZD	Algerian dinar	THB	Thai baht
EGP	Egyptian pound	TRL	Turkish lira
EUR	Euro (Europe)	UAH	Ukrainian grivna
IDR	Indonesian rupiah	USD	US dollar
INR	Indian rupees	UYU	Uruguayan peso
JPY	Japanese yen	ZAR	South African rand

Executive summary

The *Agricultural Outlook 2016-2025* is a collaborative effort of OECD and FAO. It brings together the commodity, policy and country expertise of both organisations and input from collaborating member countries to provide an assessment of medium-term prospects of national, regional and global agricultural commodity markets. The *Outlook* provides supply, demand, trade and price estimates of major agricultural commodities for 41 countries and 12 geographical regions. The special theme chapter of this year's edition focusses on the prospects and challenges of the agricultural sector in Sub-Saharan Africa.

Prices for the main crops, livestock and fish products all fell in 2015, signalling that an era of high prices is quite likely over for all sub-sectors. Meat prices fell from record highs in 2014, dairy product prices continued declines that started in 2013 and 2014, while crop prices fell further from their peaks in 2012. The main factors behind lower prices have been several years of robust supply growth, weakening demand growth due to the overall economic slowdown, lower oil prices and further accumulation of already abundant stocks.

Over the ten-year *Outlook* period the demand growth for food is expected to slow progressively. Global population growth, the main driver of demand increases, is declining, while income growth in emerging economies is projected to be weaker. At the same time, consumers, especially in populous emerging economies, show a declining propensity to spend income gains on consuming more basic foodstuffs. Demand for meat, fish and dairy products will grow relatively strongly, inducing additional demand for feed, in particular from coarse grains and protein meals. Demand for agricultural commodities for biofuel production is projected to stagnate due to the lower energy prices and more conservative biofuel policies in several countries.

Increasing consumption in developing countries is projected to reduce the global proportion of people who are undernourished from 11% to 8% over the next ten years, with the total numbers of undernourished declining from 788 million to less than 650 million. However, undernourishment in Sub-Saharan Africa remains high, and in ten years the region will account for more than one third of the global total of undernourished, compared with just over a quarter today. Many countries will be confronted with a complex burden of undernourishment (too few calories), obesity, and micronutrient deficiency (with unbalanced diets a common problem). In both developed and developing countries, consumption of sugar, oils and fats is projected to increase faster than consumption of staples and protein, largely as a result of people consuming more processed food products.

The increased demand for food is projected to be satisfied through productivity gains, with modest changes in crop area and livestock herds. Yield improvements are projected to account for 80% of the increase in crop output. There is some scope to increase agricultural area sustainably, mainly in parts of Latin America and Sub-Saharan Africa. The majority of new crop area in Africa will be dedicated to cereals, while the expansion in

Latin America is focussed on soybeans. Yield growth is expected to be slower in the main producing countries, as it becomes progressively more difficult to shift the technological frontier forward. But there are large yield gaps in many developing countries, especially in Sub-Saharan Africa, and bridging these gaps could add significantly to global supplies.

With overall market growth projected to slow, agricultural trade is expected to expand at about half the rate of the previous decade. Nevertheless, for most commodities a constant share of production continues to be traded on world markets. The fact that relatively few countries are abundant in natural resources means that trade will become more important for global food security. However, trade in basic food products is constrained by the pursuit of food self-sufficiency policies in a number of countries, and a structural shift towards more trade in value added products.

For most agricultural commodities, global exports are concentrated among just a few key supplier countries. For all products covered by the *Outlook*, the five main exporters will account for at least 70% of total exports, with just two or three countries dominating supplies of some commodities. On the import side there is less concentration, although the People's Republic of China (hereafter "China") is a critical market for some commodities – in particular soybeans, but also dairy products and coarse grains other than maize. Food import dependency of resource poor regions, especially North Africa and the Middle East, is projected to intensify.

With supply and demand growth broadly matched, real agricultural prices are projected to remain relatively flat. However, there will be some relative price changes that reflect adjustments in the composition of demand, as well as differences in supply conditions, such as the comparative ease of increasing production in Latin America relative to Asia. Overall, livestock prices are projected to rise relative to crop prices, and the prices of coarse grains and oilseeds are projected to rise relative to the prices of food staples. Those structural trends are likely to be more apparent in the current context of lower prices across all commodity groups.

The *Outlook* is subject to a wide range of uncertainties, including variations in oil prices, yields and economic growth. If historical variations in these factors continue, then there is a strong chance of at least one severe price swing within the next ten years. Such wide inter-annual price movements can mask long-term trends. Climate change may add to this uncertainty, especially if the occurrence of extreme weather events intensifies.

In addition, there are several policy uncertainties. One relates to China's recently announced changes to its grains policy, including the setting of domestic prices and the management of stocks. The current *Outlook* assumes that those changes will enable China to meet its domestic objective of maintaining a high self-sufficiency ratio in maize without severely disrupting international markets. However, the timing and scale of stock release is a major uncertainty underlying the projections. A further risk relates to the Russian import ban, which is assumed to expire at the end of 2017.

Sub-Saharan Africa

The Sub-Saharan Africa (SSA) region accounts for more than 950 million people, approximately 13% of the global population. Despite ongoing transformation of the region's economies, agriculture remains a crucial sector providing livelihoods for millions of people. Regional differences in the structure and development stage of agriculture reflect the vast agro-ecological, economic, political and cultural differences across the continent. Undernourishment has been a long-standing challenge, with uneven progress toward food security across the region.

Development of the region's agricultural sector is being shaped by rapid population growth, urbanisation and rural diversification, an associated structural transformation from farm to non-farm employment, the rise of a middle class, and increasing interest (both domestically and globally) in the continent's farmland. Total agricultural production is projected to expand by 2.6% p.a. In contrast with past production increases, which overall were driven by area expansion, an increasing share of future production growth will come from improved productivity. Inclusive development will be needed that improves the productivity of small-scale, resource poor farmer, while creating broader rural development opportunities.

Assuming continued rapid population growth across the region, complemented by rising incomes and continuation of current policies and market structures, the production of food crops in many countries is projected to grow more slowly than demand. Sub-Saharan Africa's net imports of food commodities are anticipated to grow over the next decade, although productivity enhancing investments would mitigate this trend.

Many countries are competitive producers and regular exporters of fruit and beverage crops, which contribute to foreign currency reserves. Such products may offer farmers alternative opportunities to traditional food crops. They may also be a potentially important source of employment for the continent's young population. With a limited number of food exporters, and a large number of net importers, open regional trade will be central to food security.

While the outlook for agriculture in Sub-Saharan Africa is broadly positive, it could be much improved by more stable policies across the region, by strategic public and private investments, notably in infrastructure, and by suitably adapted research and extension. Such investments could improve access to markets, reduce post-harvest losses, and make needed inputs more widely available.

PART I

Overview and special chapter

PART I

Chapter 1

Overview of the OECD-FAO Agricultural Outlook 2016-2025

This chapter provides an overview of the latest set of quantitative medium-term projections for global and national agricultural markets. The projections cover production, consumption, stocks, trade and prices for 25 agricultural products for the period 2016 to 2025. The chapter starts with a description of the state of agricultural markets in 2015 and explains the main macroeconomic and policy assumptions underlying the projections. In the next sections, consumption and production trends are examined, with a focus on regional developments. The chapter also reviews trade patterns showing the relative concentration of exports and dispersion of imports across countries for different commodities. The chapter concludes with global agricultural price projections, which include a stochastic analysis to illustrate how uncertainty about the macroeconomic environment and yield levels might affect price projections. Growing demand for agricultural commodities is projected to be matched by efficiency gains in production which will keep real agricultural prices relatively flat.

The statistical data for Israel are supplied by and under the responsibility of the relevant Israeli authorities. The use of such data by the OECD is without prejudice to the status of the Golan Heights, East Jerusalem and Israeli settlements in the West Bank under the terms of international law.
The position of the United Nations on the question of Jerusalem is contained in General Assembly Resolution 181(II) of 29 November 1947, and subsequent resolutions of the General Assembly and the Security Council concerning this question.

The setting: Year-over-year prices of crops and livestock at lower levels in 2015

Whereas 2014 witnessed a divergence in crop and livestock markets with decreasing crop prices and increasing livestock prices, prices for both crops and livestock fell in marketing year 2015 (see the glossary for a definition of marketing year). Production of most crops declined in 2015 compared to the exceptionally high output levels in 2014. However, this fall in output could not halt the decline in crop prices which was fuelled by lower demand and high inventories. Livestock prices came down from their historic highs in 2014 following weaker demand growth and lower feed grain prices. Dairy prices continued their decline that started during 2014 amid reduced import demand and higher output. An increased supply within certain fishery species combined with lower consumer demand in key markets led to generally declining fish prices.

The main factors depressing agricultural prices in 2015 were continued strong supply combined with weak economic growth and abundant stocks. Global commodity market conditions as well as macro-economic and policy developments will influence agricultural price trends in the medium term. A short overview of the current market conditions for each commodity is presented below, while Box 1.1 describes in detail the main macro-economic and policy assumptions that are adopted in the baseline projections. Growth prospects in industrialised economies are expected to average about 2% p.a., while those for major emerging economies range from India at about 7.5% p.a. to only 1.2% p.a. in the Russian Federation. Developing and least developed countries' growth prospects are generally expected in the 5-7% p.a. range. Developing countries will continue to drive population growth; however global population growth is projected to slow to 1% p.a. over the next decade. Inflation in OECD countries and the People's Republic of China (hereafter "China") will remain low, but is projected to be quite high in several emerging economies (Argentina, Brazil, and South Africa) due to the impact of sizeable currency depreciations. The oil price is projected to increase from USD 39.3 per barrel in 2016 to USD 83.2 per barrel in 2025, and will underpin rising nominal prices for agricultural commodities.

The baseline projections in the *Outlook* reflect specific assumptions regarding the factors that influence supply, demand, trade and prices. These assumptions include stable macro-economic and weather conditions. In the last section of the chapter, the sensitivity of the price projections to these assumptions is examined. This complementary analysis provides an indication of the range of possible outcomes around the baseline, given the variability of yields and macro-economic conditions, including GDP growth, oil prices, and exchange rate fluctuations.

This edition of the *Outlook* includes for the first time a disaggregation of coarse grains into maize and other coarse grains, the disaggregation of oilseeds into soybean and other oilseeds, the decomposition of aquaculture production into the main groups of species, and the development of a new specification that provides projections for consumer prices.

Current market conditions

Following the record harvests of 2014, production of wheat, maize, rice and other coarse grains declined in 2015. Despite this fall in production, international prices of all cereals remained under downward pressure due to large global inventories and slower demand growth. Prices of wheat dropped to their lowest levels since 2009, with global stocks reaching their highest level since that year.

Whereas soybean production increased in 2015, the production of other oilseeds (rapeseed, sunflower seed, and groundnuts) was lower than in 2014. Vegetable oil production slowed relative to oilseed production because of a decline in palm oil production caused by *El Niño* in South East Asia and because of the increasing share of soybeans in the global oilseeds markets. The demand for vegetable oil weakened due to a reduction in biofuel production in Indonesia. Prices for protein meal also declined relative to prices of coarse grains and other feed ingredients.

Even though sugar production in Brazil and other smaller producers increased, global sugar production decreased in 2015, following lower production in Europe and several key Asian countries. The price of sugar remained low due to a combination of huge stocks, weak economic growth and, low oil and crop prices.

In 2015, weak demand growth for meat combined with production responses to lower feed costs exerted downward pressure on prices, which fell to levels last witnessed in 2010. Weakening import demand from the Russian Federation and reduced supply from North America restrained world trade volumes. Production growth was largely driven by developing economies dominated by Brazil and China, which are the largest meat producers in the developing region. The increase in meat consumption was mainly driven by poultry, which accounted for two thirds of the additional meat consumed.

Lower import demand in China and growing production in the European Union, the United States and Oceania were the main drivers of lower dairy prices in 2015. The removal of the EU milk quota in April 2015 allowed milk production to increase, but the supply response has been uneven across member countries. The ongoing Russian Federation's import ban continues to affect butter and cheese exports from the European Union, the United States and Australia.

Developing countries dominated a sustained growth in fish consumption and production in 2015. Aquaculture remains the main motor behind the expansion in global fish supply. Economic contraction and exchange rate fluctuations in key players caused a decline in trade of fish and fishery products in value terms. Price evolutions varied between the different species and markets, but in general prices of wild species are stronger than those of farmed seafood.

Weak crude oil and feedstock prices in 2015 led to decreasing ethanol and biodiesel prices. Demand for biofuels continues to be highly influenced by domestic policies in conjunction with sustained fuel demand across the world.

Cotton production in almost all major producing countries fell considerably during the first half of 2015, leading to a drop of 9% in global production. The unexpected sharp decline was a result of adverse weather, lower global market demand and policy uncertainty. In response to the plunge in production, global stocks – which had reached record levels in 2014 – were released in major producing countries, notably China where the gap between the prices of domestic and imported cotton has narrowed in response to the elimination of minimum support prices in 2014. As a consequence cotton trade declined further.

The new Sustainable Development Goals (SDGs) include a significant number of interconnected objectives related to agriculture and food. SDG 2 focuses explicitly on food

Box 1.1. **Macro-economic and policy assumptions**

The main assumptions underlying the baseline projection

The *Outlook* is a base scenario that is considered plausible given a range of conditioning assumptions. These assumptions present a specific macro-economic, policy, and demographic environment which underpins the projections for the evolution of demand and supply for agricultural and fish products. This environment is described below.

Slowing global economic activity

Global growth continues to underperform expectations and in 2015 eased to around 3%, well below its long-run average. This largely reflects further weakness in emerging market economies (EMEs). Deep recessions have emerged in Brazil and the Russian Federation, while the ongoing slowdown in the People's Republic of China (hereafter "China"), and the associated weakness of commodity prices has hit activities in key trading markets and commodity exporting economies. Increased financial market uncertainty and the slowdown of global trade growth, especially in EMEs, are other headwinds to global activity.

Growth in the OECD economies as a group remained consistent in 2015 at around 2%, although growth slowed for some countries, including Australia, Canada, Korea and New Zealand. On the other hand, growth in Chile, Mexico, Turkey, Japan, and EU15 members as a group improved in 2015. The United States and Israel maintained the same level of growth in 2015 as in 2014.

The macro-economic assumptions used in the *Agricultural Outlook* are based on the *OECD Economic Outlook* (November 2015) and the International Monetary Fund's, *World Economic Outlook* (October 2015).

Relative to 2015, growth is expected to pick up slightly in advanced economies, but to decline in emerging market and developing economies. Supportive macro-economic policies, lower commodity prices and a further steady improvement in labour market outcomes should continue to support the recovery in the advanced economies, with GDP growth projected to average 2.1% per annum (p.a.) until the end of the projection period for the OECD countries as a group.

Among OECD countries, Korea and Turkey are expected to be the growth leaders over the next ten years, with average annual rates of 3.5%, followed by Mexico at 3.2%. The recovery is expected to continue in the United States, supported by lower energy prices, reduced public deficits, and an improving housing market. Medium-term growth in the United States is expected to remain relatively solid at an annual rate of 2.4% over the next ten years. The moderate recovery in the euro area is projected to continue in 2016, sustained by lower oil prices, monetary easing (reflecting subdued inflation), and depreciation of the euro. The outlook is for an annual average growth rate of 1.7% for EU15 members as a group during the projection period.

Growth prospects are likely to continue to diverge in the large EMEs. A gradual slowdown is projected to continue in China, with GDP growth easing to 6.2% p.a. over the next ten years compared to an annual average growth rate of 9.0% during the last decade. Growth prospects in India should remain relatively robust at 7.6% p.a., provided further progress is made in implementing structural reforms. Despite large currency depreciations, recovery will be only gradual in Brazil and the Russian Federation, occurring as of 2017 and averaging 2% p.a. and 1.2% p.a. respectively until the end of the projection period.

Growth prospects for developing economies are still expected to remain strong. Most developing countries are expected to achieve marginally lower growth than in the previous decade. Bangladesh and the Philippines are expected to have the leading growth in Asia, averaging 6.6% p.a. and 6.3% p.a. respectively. Growth in Sub-Saharan Africa is anticipated to slow to about 4.7% p.a. under lower commodity prices. Underpinned by improvement in macro-economic stability and sectoral reforms supporting private investment, Ethiopia and Mozambique are expected to have the strongest growth over the coming decade, growing respectively at rates of 7.2% p.a. and 7.6% p.a., while Tanzania should maintain an annual average growth of 6.7%. Countries in North Africa and the Middle East are expected to grow 3.5% p.a., down considerably from the 5.3% p.a. average of the past decade. Latin America is set to keep growing at similar rate as in the last decade, with an annual average growth rate of 3% during the projection period, somewhat slower than the averages for Asia and Sub-Saharan Africa.

Box 1.1. **Macro-economic and policy assumptions** *(cont.)*

Measures of income growth are assumed to apply across the entire population and influence the demand for various food products, but in reality economic growth may not be spread so equitably and consumers at the low end of the income distribution may not see a corresponding increase in their incomes. Furthermore, while many developing and least developed economics are growing strongly, they do so from a very low base and the absolute increase in incomes remains quite small. These factors explain why, despite long periods of strong growth, food consumption patterns have changed relatively slowly.

Figure 1.1. **GDP growth rates in OECD and developing countries**

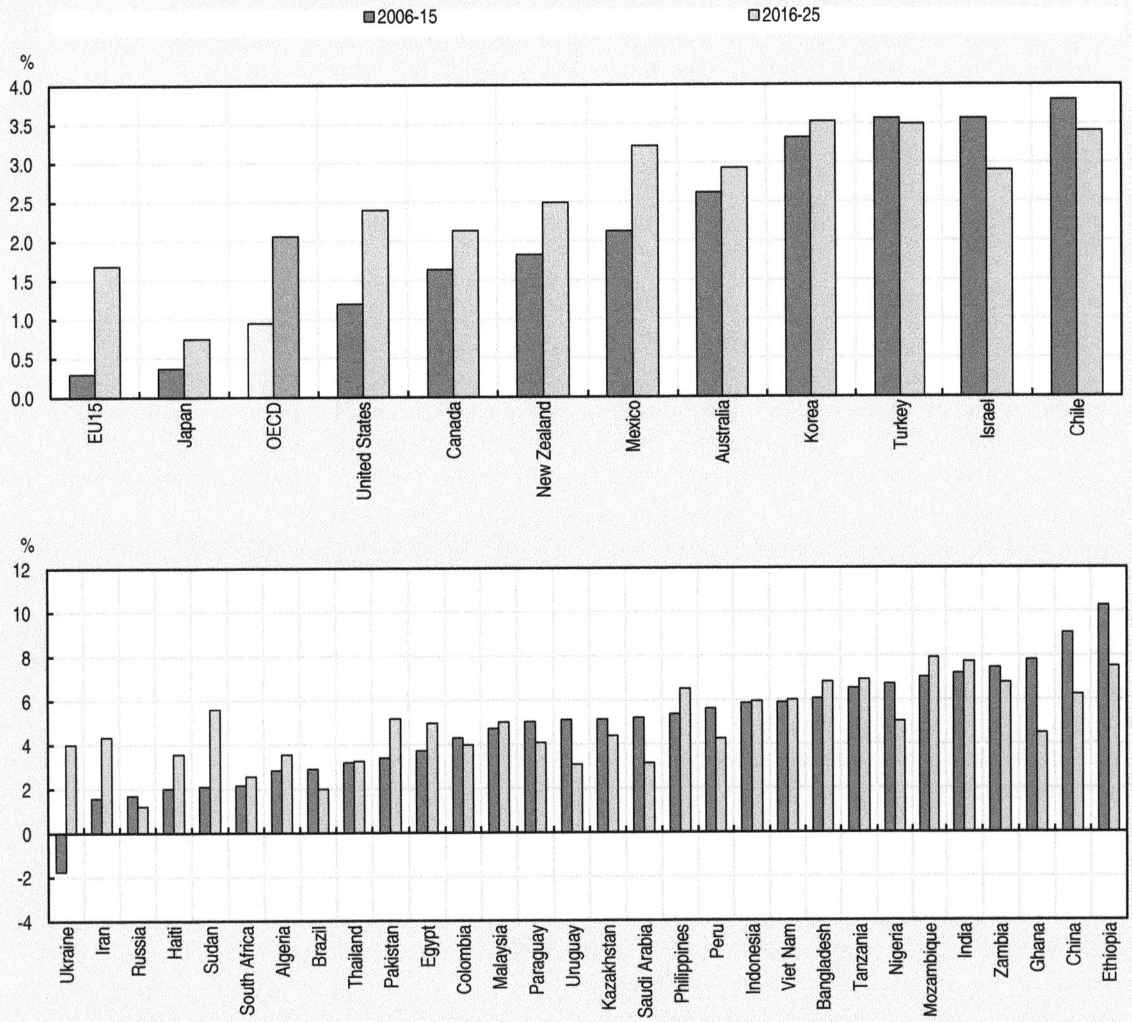

Source: OECD/FAO (2016), "OECD-FAO Agricultural Outlook", *OECD Agriculture statistics* (database), *http://dx.doi.org/10.1787/agr-data-en.*

StatLink *http://dx.doi.org/10.1787/888933381153*

Slowdown in population growth

World population growth is expected to slow to 1% p.a. over the next decade. Population growth is still driven by developing countries, particularly Africa which is expected to exhibit the fastest growth rate at 2.5% p.a.; some countries of Africa will witness population growth in excess of 3.5% p.a. The Asia and Pacific region will account for nearly half of the world's population, and India, with an additional 151 million people by 2025, should overtake China as the most populous country.

Box 1.1. **Macro-economic and policy assumptions** *(cont.)*

Among the OECD countries, the Japanese population is expected to decrease by 3.7 million during the next ten years and the population of the Russian Federation will also decrease by 2.3 million. The population of the European Union is expected to remain stagnant, growing at a rate of 0.08% p.a.; Australia has the highest projected population growth among OECD countries at 1.18% p.a., followed by Mexico at 1.10% p.a.

Inflation

Inflation remains weak in the OECD area and in China, but has picked up in several other EMEs, particularly those where large currency depreciations have occurred. Inflation is close to zero in Japan, the United States and the European Union, but is projected to rise in 2016 and thereafter. Inflation in these countries, however, is generally expected to remain below central bank targets.

Consumer price inflation should remain relatively low in China and India, helped by a weak import price pressure. Among other major EMEs, inflation is projected to remain strong for several years to come in countries such as Argentina, Brazil, the Russian Federation, and South Africa due to the impact of significant currency depreciations, and the domestic effect of sanctions in the case of the Russian Federation.

Recent currency movements have been unusually large

In nominal terms, the euro depreciated by 19% during the last three years relative to the US dollar and by 25% relative to the Japanese yen. Large exchange rate changes have also occurred for a number of emerging market and developing economies. Weakening commodity prices are reflected in the sizable currency depreciation experienced by many commodity exporters with flexible exchange rate regimes. But large exchange rate movements have not been limited to commodity-exporting emerging markets as currencies of EMEs more generally have seen sharp depreciations.

Argentina's peso re-entered the international currency market in December 2015, leading to an immediate and large depreciation.

Nominal exchange rates for the period 2016-25 are mostly driven by the inflation differential in relation to the United States (with small changes in real terms). Given the expected inflation differentials, currencies will appreciate in nominal terms relative to the US dollar over the next ten years in Japan, Canada, Korea, the Euro area, China, and the Russian Federation. Conversely, a very strong depreciation of Argentina's, Brazil's, India's, South Africa's and Turkey's currencies is projected over the next decade.

Collapse of energy prices

Oil prices have dropped sharply since mid-2014, driven by a slowing demand and record increases in supplies, particularly shale oil from North America, as well as the decision by the Organization of Petroleum Exporting Countries (OPEC) to leave its production target unchanged.

World oil price assumptions to 2014 are obtained from the short-term update of the *OECD Economic Outlook N°98* (November 2015). For 2015, the annual average daily spot price is used, while the average daily spot price for December 2015 is used as the oil price value for the year 2016. Oil prices during the projection period are expected to grow at rates projected by the *World Energy Outlook* (IEA, November 2015).

Prices remain lower for much of the early part of the projection period, until the market rebalances at higher price levels via higher demand and lower growth in supply. In nominal terms, oil prices are expected to increase over the outlook period at an average annual rate of 8.3%, from USD 39.3 per barrel in 2016 to USD 83.2 per barrel by 2025.

Box 1.1. **Macro-economic and policy assumptions** *(cont.)*

Policy considerations

Policies play an important role in agricultural, biofuel and fisheries markets, with policy reforms often changing the structure of markets. This *Outlook* assumes that policies will remain as they are throughout the projection period. In the case of bilateral trade agreements, only ratified agreements are incorporated. For example, the ratification process of the Trans-Pacific Partnership (TPP) is ongoing; consequently the TPP is not incorporated into this *Outlook*. The Russian Federation's ban on imports from specific origins was announced as a temporary measure that would remain in place until 2017 and is incorporated as such in the *Outlook*.

by seeking to "end hunger, achieve food security and improved nutrition and promote sustainable agriculture", but multiple other goals relate to challenges in the food system. SDG 1 focuses on poverty reduction, where agriculture and food have a key role to play. Sustainable agriculture plays a central role in achieving SDG 6 on water, SDG 12 on sustainable consumption and production, SDG 13 on climate change adaptation and mitigation and, SDG 15 on land use and ecosystems. Some of the principal linkages between agriculture and the SDGs are summarised in Box 1.2. The current baseline, which assumes that current policies remain in place, does not incorporate the effects of policy efforts to address these goals. However, there is an examination of the implications of the 2015 baseline projections for hunger, as captured by the FAO's undernourishment indicator, and hence for progress towards SDG 2.

Box 1.2. **How will the United Nations Sustainable Development Goals affect agriculture?**

On 1 January 2016, the United Nations' 17 Sustainable Development Goals took effect, launching the countdown towards the achievement of 169 targets by 2030, and even by 2020, in some cases. Many of these ambitious targets are deeply significant for agriculture.

Of foremost importance is the second Goal, **End hunger, achieve food security and improved nutrition and promote sustainable agriculture**, which features targets on hunger, malnutrition, productivity and incomes, sustainability and resilience, biodiversity, investment, trade and commodity markets. In view of the close connections which exist between hunger, malnutrition, agricultural productivity and *poverty* worldwide, however, the first Goal, **End poverty in all its forms everywhere**, is also highly relevant. While food security has many facets, the majority of the world's hungry are *chronically* hungry, and they are chronically hungry because they are poor. Given that 78% of the world's poor are heavily dependent on agriculture not only for their food, but also for their livelihoods, agricultural development, including the growth of agricultural productivity and incomes, represents one of the most powerful tools that exists to end extreme poverty and feed 9 bln people by 2050 (World Bank, 2015).

Beyond the first two Goals, the majority of the SDGs, listed below, are either directly or indirectly relevant for agriculture. Together, these supplement and reinforce the message previously put forward by the UN Millennium Development Goals, namely that business as usual practices cannot continue.

Abbreviated goal titles and relevance for agriculture*

1 No poverty: Includes targets for the eradication of extreme poverty (incomes of less than USD 1.25 a day) and at least 50% reduction of poverty by 2030. Reference is also made to ownership and control over land and natural resources. This target is relevant for agriculture given its importance for poverty reduction in many developing countries in particular.

Box 1.2. **How will the United Nations Sustainable Development Goals affect agriculture?** *(cont.)*

2 Zero hunger: Numerous relevant targets, including the ending of hunger and malnutrition; the doubling of agricultural productivity and incomes of small-scale food producers; the correction of international trade restrictions; increased investment in agricultural research, extension services and technology; and the implementation of sustainable food production systems and practices by 2030.

3 Good health and well-being: Includes the reduction of deaths and illnesses from hazardous chemicals and air, water and soil pollution and contamination. Relevant for agro-chemical and manure use.

4 Quality education: Relevant for agricultural training and extension, both of which can enable farmers to adopt environmentally sustainable farming practices and improve competitiveness.

5 Gender equality: Includes the eradication of gender discrimination, including in land ownership. Bridging the yield gap which currently exists between male and female farmers in developing countries could increase total agricultural output of these countries by 2.5-4%, thereby reducing global hunger by 12-17%.[1]

6 Clean water and sanitation: Relevant for water use efficiency of agricultural production, the improvement of water quality via the reduction of pollution, and the protection of water-related ecosystems. Farming accounts for around 70% of water used worldwide and contributes to water pollution from excess nutrients, pesticides and other pollutants.

7 Affordable and clean energy: Includes targets for the substantial increase of renewable energy and doubling of the improvement in global energy efficiency by 2030. Relevant for agricultural energy use efficiency and bioenergy production.

8 Decent work and economic growth: Features relevant targets for sustainable per capita economic growth, improvement of resource use efficiency, and access to financial services and insurance.

9 Industry, innovation and infrastructure: Agriculture-relevant targets include the development of sustainable and resilient infrastructure, increased SME access to financial services and their integration into value chains, and the encouragement of innovation.

10 Reduced inequalities: Targets include the achievement, by 2030, of the sustainable income growth of the bottom 40% of the population at a higher rate than the national average. Relevant for reasons mentioned under Goal 1.

12 Responsible production and consumption: Includes the reduction of food losses and waste, efficient use of natural resources, environmentally-sound management of chemicals and waste, the reduction of fossil-fuel subsidies, and the ten-year framework on sustainable production and consumption.

13 Climate action: Targets strengthened resilience to climate-related hazards and the joint mobilisation of USD 100 bln annually by 2020 to facilitate climate change mitigation by developing countries. Climate change is highly relevant for agriculture, both impacting and being impacted by it.

14 Life below water: Includes the prevention and significant reduction by 2025 of marine pollution, nutrient pollution in particular; the effective regulation of fishing to ensure sustainable fishing practices; and the prohibition of certain fisheries subsidies by 2020.

15 Life on land: Targets the conservation and sustainable use of freshwater ecosystems and their services – such as wetlands – and the promotion of sustainable forest management – the halting of deforestation included – by 2020, the combatting of desertification and the restoration of degraded land and soil by 2030, and the prevention of biodiversity loss. As farming is the human activity which occupies the largest share of total land in many countries, agriculture can not only significantly impact biodiversity, but is also dependent upon it.

17 Partnerships for the goals: Features agriculture-relevant targets on international trade, including the promotion of an open, non-discriminatory and equitable multilateral trading system and the conclusion of the WTO Doha Development Round.

1. FAO (2011), *The State of Food and Agriculture 2010-11*: *www.fao.org/publications/sofa/2010-11/en/*

Abbreviated titles as per United Nations formulation. For full titles, please refer to *https://sustainabledevelopment.un.org*.
Goals 11 (Sustainable cities and communities) and 16 (Peace, justice and strong institutions) excluded from list.

Consumption

Population growth in developing countries is the dominant driver of global food consumption

The main driver of global agricultural commodity demand over the next decade will be population growth in developing countries. The world's population is projected to rise from 7.4 bln in 2016 to 8.1 bln in 2025, with 95% of this increase occurring in developing countries. By 2025, there will be 6.7 bln people living in developing countries and 1.4 bln in developed countries. This means that between 2005 and 2025 the population in the developing countries will have expanded by an amount that equals the entire population in developed countries. The fastest growth in population will occur in Sub-Saharan Africa (SSA), with an increase from 0.96 to 1.22 bln, or 2.7% p.a., between 2016 and 2025.

A second determinant of consumption demand is per capita income growth, which adds to the consumption of each person. Again the main impetus for consumption growth will come from developing countries since their per capita income growth rate is expected to be higher. In addition, poorer people tend to spend a greater share of their additional income on food (i.e. their income elasticity of demand for food is higher). For example, the share of additional income that will be spent on food in China and the United States in 2025 is projected to correspond to 3.4% of the income increase in China and only 1.1% of the income increase in the United States.

Rising per capita incomes are associated with a third factor: changing consumer habits. As countries develop, they pass through a "nutrition transition", by which higher incomes translate first into a demand for more calories, and then into a demand for more protein (typically from animal sources) as well as for other nutrients coming from fruit and vegetables. This trend is accompanied by more consumption of sugar, oils and fats, and greater consumption of processed foods. Many developing countries have complex consumption structures, with people who are undernourished (not having enough calories), over-nourished (having too many) and malnourished (often as a result of consuming an unhealthy diet). Developed countries mostly have more stable food preferences, with incomes growing more slowly and consumption patterns less sensitive to income changes. There are rising levels of over-consumption and a particular tendency to consume more meat and dairy products, as well as vegetable oils and sweeteners.

Per capita food consumption dominated by cereals in developing countries

Population growth, per capita income growth and changing consumer habits are integrated into the projections for per capita food consumption. Figure 1.2 shows the differences in per capita food demand between developed and developing countries, with the latter broken down into Sub-Saharan Africa and "other developing countries". The majority of least developed economies (LDCs) are located in Sub-Saharan Africa. South Africa, which is traditionally included in the group of developed countries, is considered as part of Sub-Saharan Africa for consistency with Chapter 2.

The panel on the left shows projected per capita consumption in 2025 of the main food items covered in the *Outlook*, while the panel on the right presents their respective growth rates between 2013-15 and 2025. These figures only consider food use of the commodities that are covered in the *Outlook* and hence exclude some important ingredients in people's diets, in particular vegetables, fruit and pulses.

Figure 1.2. **Per capita food consumption by region**

Kg/cap/year in 2025 (left) and growth 2025 vs. 2013-15 (right)

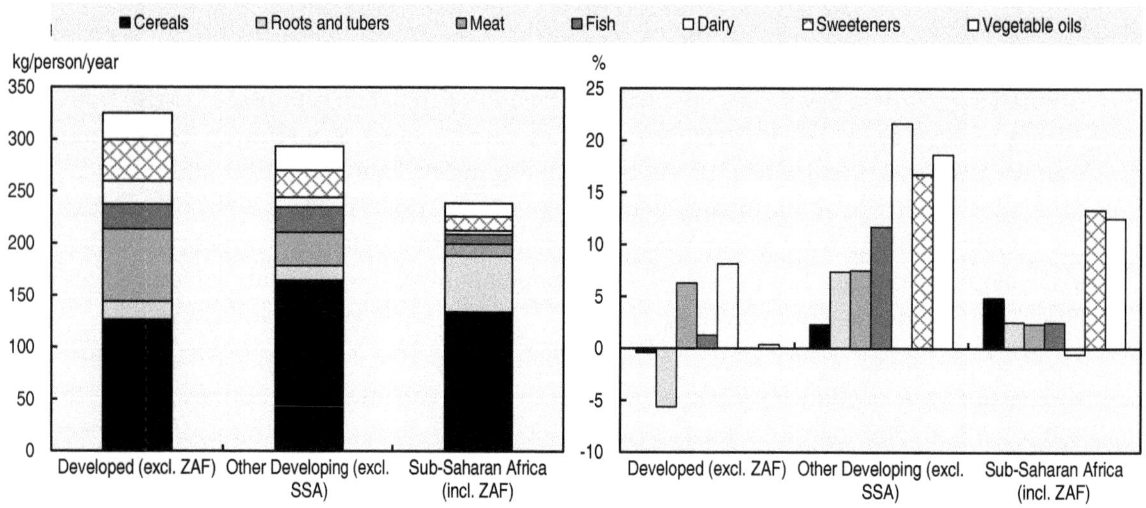

Note: Dairy products are represented in dry equivalent because their high water contents would lead to a disproportionate per capita consumption compared to the other food items. ZAF refers to South Africa.

Source: OECD/FAO (2016), "OECD-FAO Agricultural Outlook", *OECD Agriculture statistics* (database), *http://dx.doi.org/10.1787/agr-data-en.*

StatLink *http://dx.doi.org/10.1787/888933381166*

In 2025, developed countries will still consume the highest volume of food in per capita terms. However, the gap between developed and developing countries (excluding SSA) is narrowing and per capita consumption in some developing countries has already exceeded the developed average. Per capita consumption in SSA on the other hand will still be almost 20% lower than in other developing countries, with about half of all calories coming from cereal consumption.

In all three country groups, cereals represent the main ingredient in the diets in per capita terms but their relative importance is decreasing slightly over the projection period. Consumers in developed countries will continue to reduce their cereal intake and switch to other sources of energy. Per capita cereal consumption for food purposes will grow only slightly in developing countries (excluding SSA), but is projected to increase by 4.9% in SSA. Interestingly, rice consumption in SSA will experience the largest growth rate among the cereals (8.3%), growing from 25.8 to 27.9 kg/cap/year between the base period and 2025. Roots and tubers, in particular cassava, remain an important component of diets in SSA, with per capita consumption of 53 kg in 2025.

Per capita consumption of meat in 2025 is estimated at 69.7 kg retail weight (rwt) in developed countries; which is more than twice that in the other developing countries (32 kg rwt), and almost seven times that in SSA (11.3 kg rwt). In developed countries, meat exhibits a strong growth mainly as a result of solid demand growth in North America, which is fuelled by the strength of the US economy under the baseline and by lower prices. Per capita consumption of fish in 2025 is expected to remain lower in developing countries than in developed countries (21.5 kg live weigh (lw) vs 23.3 kg lw). However, if SSA is excluded, per capita fish consumption in 2025 in developing countries (excluding SSA) would be 24.3 kg lw and surpasses consumption in developed countries.

Per capita consumption of dairy products in developing countries (excluding SSA) will increase by 21% compared to the base period and be heavily oriented towards fresh dairy products over processed dairy products. The largest per capita consumption increases of fresh dairy products will be observed in India, Pakistan, Turkey and Uruguay. In developed countries, per capita consumption is projected to grow even faster in the next decade than in the previous one. Demand growth will be predominantly driven by consumers in Ukraine and the Russian Federation, where the import ban is expected to be lifted as of 2017. SSA is expected to witness only a marginal increase in per capita dairy consumption.

In overall terms, the consumption of more meat, fish and dairy products will lead to more diversified diets and higher per capita protein intake. On a global scale, the tendency of meat consumption to rise with incomes outweighs tendencies to reduce meat consumption in countries where per capita consumption levels are already high.

Developing countries exhibit strong per capita food demand for sugar and vegetable oil

In developing countries human sugar consumption will increase more than 15% in per capita terms. This translates into increases from 20 kg to 23 kg per capita in developing countries (excluding SSA) between the base period and 2025 and from 11 kg to 12 kg in SSA. A large variation in sugar consumption among developing countries in 2025 is expected, ranging from 2 kg in least developed Oceania to above 50 kg in Uruguay (52 kg), Brazil (57 kg), Thailand (56 kg) and Malaysia (65 kg). In order to help countries identify the amount of sugar that is consistent with a healthy diet, the WHO recommended in 2015 that daily per capita intake of sugar should not exceed 10% of total energy intake, implying that the expected sugar consumption is proportionate to total calorie intake. Implementing the WHO guidelines will not only have impacts on the demand side but also on the production side. Box 3.3.1 in the (online) sugar chapter examines how reducing sugar consumption in line with the WHO guidelines would impact agricultural commodity markets.

Per capita consumption of sweeteners in developed countries is projected to increase only marginally by 2025. In some countries of the European Union a substitution of sugar with HFCS is expected to take place after the sugar and HFCS quota abolition. Consequently, the share of HFCS in EU sweetener consumption is expected to reach 11% in 2025, from 3% during the base period. Some countries have also started targeting obesity via market interventions. Mexico, for example, introduced an 8% tax on sugary drinks in 2014, which has been incorporated in the *Outlook* projections.

The demand for vegetable oil for human consumption in developing countries will expand considerably during the outlook period, but will remain at levels below those in the developed countries. By 2025, developing countries (excluding SSA) will consume 23.5 kg per capita, which is almost as much as in developed countries (25.5 kg), while consumption in SSA will increase to 12.8 kg. As incomes grow, vegetable oil consumption also rises. India and Thailand are projected to witness particularly high per capita growth rates of vegetable oil for food consumption, of 55% and 49% respectively.

Rising levels of food availability and increasing per capita food consumption are associated with a range of nutritional changes, some positive, some negative. Increased calorie consumption is expected to reduce undernourishment across the developing world. The second Sustainable Development Goal (SDG 2) pledges to eliminate hunger by 2030. The implications of the 2015 baseline projections for hunger, as captured by the FAO's undernourishment indicator, and hence for progress towards SDG 2, are presented in Box 1.3.

Box 1.3. **The implications of agricultural trade and market developments for food security**

Reducing hunger and undernourishment is a global priority. The Millennium Development Goals (MDGs) included a target of halving between 1990 and 2015 the proportion of people who are undernourished. According to FAO estimates, this target was almost met at the global level, but progress was uneven across countries and there remained nearly 800 million undernourished when the MDGs expired in 2015. The new Sustainable Development Goals (SDGs), which replace the MDGs, have the ambitious target of eradicating hunger entirely by 2030.

Analysis using Aglink-Cosimo provides projections for undernourishment to 2024 consistent with the baseline of the 2015 OECD-FAO Agricultural Outlook. It assesses which countries are on trend to meet the SDG objective of eliminating hunger, as captured by the FAO's undernourishment indicator. It also considers the impact on these projected numbers of four alternative scenarios; faster income growth relative to the baseline in developing countries, stronger growth in agricultural productivity, a combination of a faster income growth with a stronger productivity growth and finally a more equitable access to available food supplies.

The OECD-FAO Agricultural Outlook contains projections for production, consumption and trade of the major agricultural commodities, and implicitly for the availability of calories at the national level for the 32 developing countries explicitly covered by the Aglink-Cosimo model (OECD, 2015). The FAO's undernourishment indicator measures the probability that an individual from a reference population consumes less than the minimum calorie requirement for an active and healthy life. The Prevalence of Undernourishment (PoU) converts national calorie availability into estimates of undernourishment on the basis of an estimated distribution of peoples' access to available calories. This paper takes the *Outlook* estimates for calorie availability and – in the base case – keeps the distribution of access to those calories unchanged. On that basis it is possible to provide projections for undernourishment.

Under a "status quo" scenario in which policies remain unchanged and agricultural productivity growth continues on trend, the global PoU is projected to fall from 11% to 8% over ten years, with Latin America as a whole dipping under the 5% threshold at which the FAO considers hunger to be effectively eradicated. The PoU falls from 12% to 8% in Asia and Pacific, with Indonesia and Thailand dipping under the 5% threshold, and from 23% to 19% in Sub-Saharan Africa. The global total of undernourished people declines from 788 million to 636 million (Figure 1.3).

Figure 1.3. **Number of undernourished in selected regions**

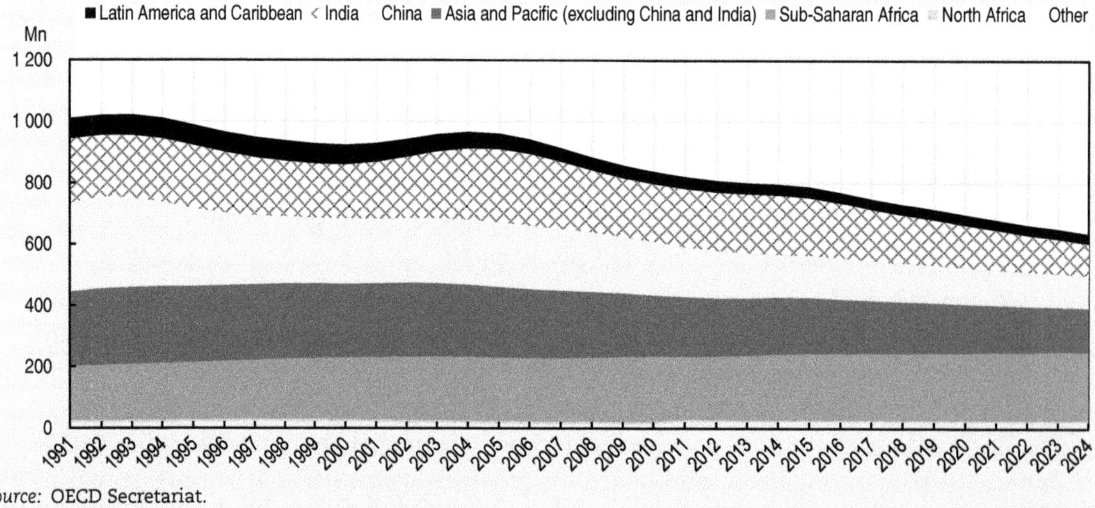

Source: OECD Secretariat.

StatLink http://dx.doi.org/10.1787/888933381170

Box 1.3. **The implications of agricultural trade and market developments for food security** *(cont.)*

By 2024 global consumption of calories from crop and livestock products will be 14% and 15% higher than in 2015 respectively. Developing countries account for 96% of the additional consumption of crop products and 88% of livestock products. Adjusting for population growth, per capita consumption levels for crops and livestock will rise by 4% and 5% respectively.

Faster income growth in developing countries of 1% cumulatively over the next ten years (raising the average annual growth rate by around a quarter) implies that per capita incomes end up 10% higher in 2024 and reduces the PoU in developing countries by a further 0.5%, with slightly larger effects in Africa than in Asia. Raising agricultural productivity growth in developing countries by 1% cumulatively over ten years (implying a gain of more than 50% across crop products relative to the baseline), would lower the PoU in developing countries by an additional 0.8% and enable China, Nigeria and Peru to effectively eradicate hunger. Here, however, the average effects would be larger in Asia than in Africa. A combination of both effects would be essentially additive, with Bangladesh joining the group of countries free from hunger.

Unsurprisingly, the biggest impacts on undernourishment come through a scenario which improves access to available calories through a more equal distribution of incomes and hence expenditures across national populations. This emphasises the fact access to food and not availability is the crucial aspect of food security since globally 50% more calories are currently available than needed to meet everyone's minimum calorie requirement. A 10% reduction in the coefficient of variation in 2024 lowers the overall PoU by 1.7 percentage points and allows all the countries in the income and productivity scenarios, plus India and Viet Nam, to eliminate undernourishment.

A combination of income growth, agricultural productivity gains, and reduced income inequality will keep most Asian countries on track to achieve the SDG of eliminating hunger. However, the PoU for Sub-Saharan Africa as a whole remains stubbornly high under all scenarios, and by 2024 the region will account for more than one third of the global total of undernourished. For the poorest African countries, much deeper transformations will be needed that raise the incomes of the poorest households and with it their access to food – whether that food is sourced domestically or from imports. The question of how to realise those gains, and the role of domestic agricultural production, is a broader development question beyond the remit of this report.

Food availability can be met by domestic production and by imports, and trade will play an important role in moving supplies from surplus to deficit countries. The importance of trade depends on each country's circumstances. Although in several cases much of the additional consumption is sourced locally many developing countries will also import a significant and rising share of total calories. The tendency will be reinforced if overall income growth is the dominant source of increased calorie availability, and reduced if the main driver is domestic agricultural productivity growth.

Overall consumption of agricultural commodities will grow faster in developing countries

Increases in per capita food consumption combine with population growth to give overall increases in consumption (Figure 1.4). A share of cereals and oilseed utilisation is destined for animal feed and the production of biofuels, so the fact that total consumption growth exceeds population growth does not necessarily imply rising levels of per capita food consumption.

Among developed countries, the United States and the European Union are the largest drivers of biofuel demand. Lower petroleum prices are stimulating gasoline use, with biofuel mandates raising biofuel use in the United States in the first two years of the *Outlook*. Given lower gasoline use prospects over the rest of the projection period and the

Figure 1.4. **Consumption growth by region**

2025 vs. 2013-15

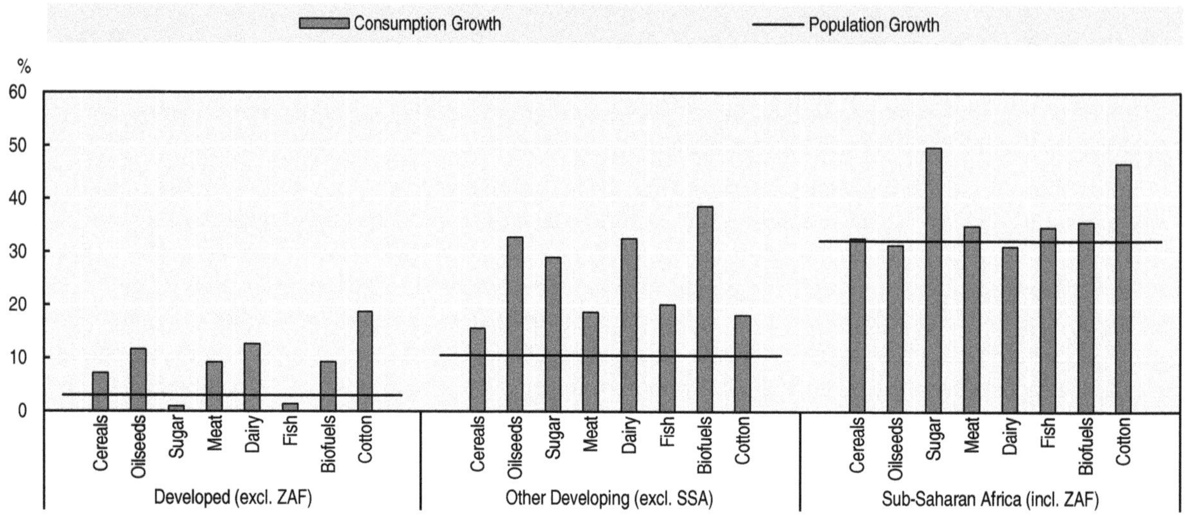

Note: ZAF refers to South Africa.

Source: OECD/FAO (2016), "OECD-FAO Agricultural Outlook", *OECD Agriculture statistics* (database), *http://dx.doi.org/10.1787/agr-data-en.*

StatLink *http://dx.doi.org/10.1787/888933381185*

restricted availability of blends exceeding the 10% blend wall, the implied maize-based ethanol mandate is assumed to decline after 2018. The advanced mandate (which covers biofuels from sources other than maize) is in turn assumed to increase so that the total US mandate remains close to its 2017 level. This translates in lower ethanol use and in an expansion of biodiesel use over the projection period. In the European Union, ethanol and biodiesel fuel use is expected to expand up to 2020, when the Renewable Energy Directive target has to be met, and then to decrease in line with lower gasoline and diesel use prospects.

Developing countries (excluding SSA) exhibit strong consumption growth rates for all commodity groups. This group of countries comprises the most populous countries as well as emerging economies which are projected to experience the largest economic and income growth. The strong demand growth for sugar occurs in populous countries in Asia and Pacific, which will account for around 67% of the expansion in use. India, China and Indonesia will experience the largest increases in sugar consumption. Brazil is projected to expand its use of sugarcane-based ethanol over the outlook period and remain the largest user of ethanol among developing countries. Biodiesel use will also expand steadily, with notable demand increases in Indonesia, Brazil and Argentina, in line with their domestic mandates. Raw cotton demand will remain concentrated in countries in Asia and Pacific. By 2025, India will have surpassed China as the world's largest consumer of raw cotton. Large cotton consumption growth rates are projected to occur in Bangladesh (3.7% p.a.), Indonesia (3.2% p.a.), and Viet Nam (3.1% p.a.).

Consumption growth in SSA is higher for most commodities than in other developing countries. However, these growth rates have to be placed in perspective, as current levels of consumption are much lower in SSA. Moreover, consumption growth in SSA for most agricultural products is only marginally ahead of population growth. Indeed, the population of SSA is expected to grow by 33.6% between 2013-15 and 2025, a much higher rate than the average 10.5% population growth in other developing countries.

Production

Production growth through intensification

Strong global demand and low stocks resulted in high agricultural prices in recent years which, along with policy reforms in many countries, provided the economic incentives and facilitating conditions for increased global agricultural production. Overall, the sector was able to expand at 2.5% p.a. in the last decade.

In the coming decade, agricultural production faces a different set of challenges. Commodity prices started to decline in 2013, stocks have been replenished and economic growth in major producer countries is projected to slow. As outlined in the previous section, global demand growth varies between commodities, but overall is projected to be slower than in the previous decade. The resulting weakening of agricultural markets makes the sector less attractive for investments, limiting total agricultural output growth to 1.6% p.a. on average during the projection period. After the strong gains in recent years, crop production is projected to grow at about 1.5% p.a. in the next decade.

Global demand expansions can be satisfied mainly through efficiency improvements, requiring only small expansions of the production base, crop area and livestock herds. In the crop sector, yield improvements will be responsible for 80% of total output growth, while area expansion accounts for 20%. Agricultural area harvested of the crops reflected in the *Outlook* is projected to expand by 42 Mha (4%) over the next decade with almost half of this increase taking place in Brazil and Argentina. Sustained demand and lower feed prices support the steady growth of livestock production at 1.4% p.a., adding about 4.5 Mt of meats and 16 Mt of milk annually. World fishery production will expand by 17% over the course of the projection period, with aquaculture production steadily increasing its role in total fish supply. Global biofuel production stagnates due to the lower energy prices, which curb discretionary demand, and more conservative policies in some of the major producer countries.

Figure 1.5. **Area and yield by region**

Area and yield growth 2025 vs. 2013-15(left) and share of agricultural area in 2025 (right)

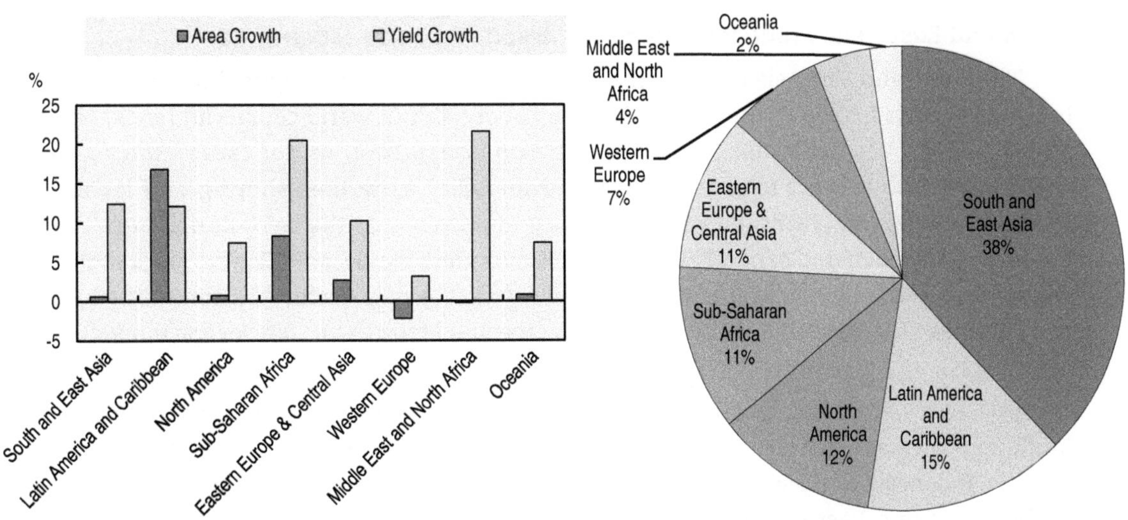

Source: OECD/FAO (2016), "OECD-FAO Agricultural Outlook", *OECD Agriculture statistics* (database), *http://dx.doi.org/10.1787/agr-data-en.*

StatLink http://dx.doi.org/10.1787/888933381196

Figure 1.5 illustrates the evolution of area growth and yield growth by geographic region and indicates each region's share of the global agricultural area in 2025 (see the methodology section for the classification of geographic regions). In Sub-Saharan Africa, production results from a combination of yield improvements and area expansion. South and East Asia, and Eastern Europe sustain their moderate growth mainly by yield advances. By contrast, in Latin America and the Caribbean area expansion is a major driver, mainly due to Brazil's and Argentina's expansion in soybean. In North Africa and the Middle East, and Western Europe, yield growth compensates for the decrease in area. In Oceania, higher self-sufficiency rates are expected through yield and area increases. The remainder of this section will provide an overview of the projections for agricultural production in the various geographic regions.

Figure 1.6. **Global meat production**

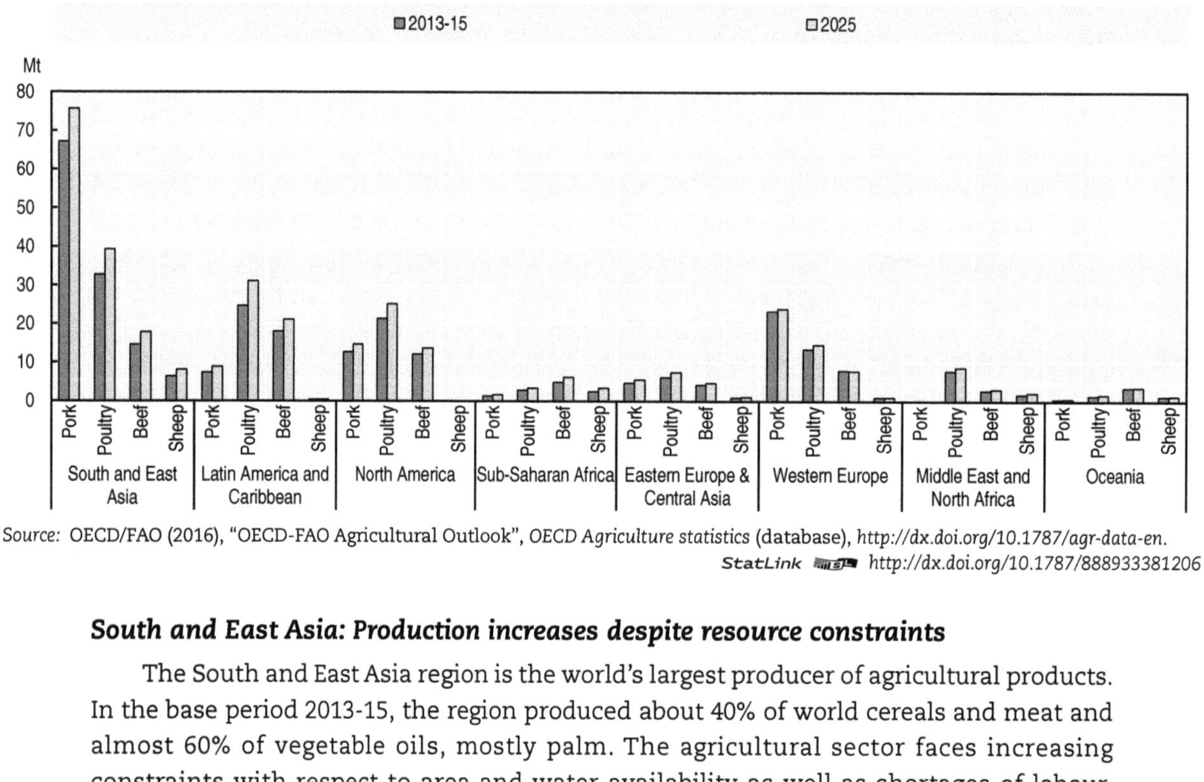

Source: OECD/FAO (2016), "OECD-FAO Agricultural Outlook", *OECD Agriculture statistics* (database), *http://dx.doi.org/10.1787/agr-data-en.*

StatLink *http://dx.doi.org/10.1787/888933381206*

South and East Asia: Production increases despite resource constraints

The South and East Asia region is the world's largest producer of agricultural products. In the base period 2013-15, the region produced about 40% of world cereals and meat and almost 60% of vegetable oils, mostly palm. The agricultural sector faces increasing constraints with respect to area and water availability as well as shortages of labour. Primarily through intensification and efficiency improvements the agricultural output in the region is expected to expand by almost 20% over the next decade. The livestock sector will contribute about 40% to this development, while the crop sector will contribute 33% and fisheries 27%. Figure 1.7 illustrates the average yield growth by 2025 for the individual crops.

Due to its dominant position and strong yield improvements, 89% of the global production increase in rice will originate in South and East Asian countries, predominately in India, Indonesia and Bangladesh. Production of maize, the second most important crop in the South and East Asia region will expand mostly on the account of China. India, currently the world's third largest wheat producer, shapes the development of this crop in

Figure 1.7. **Crop yields in South and East Asia**

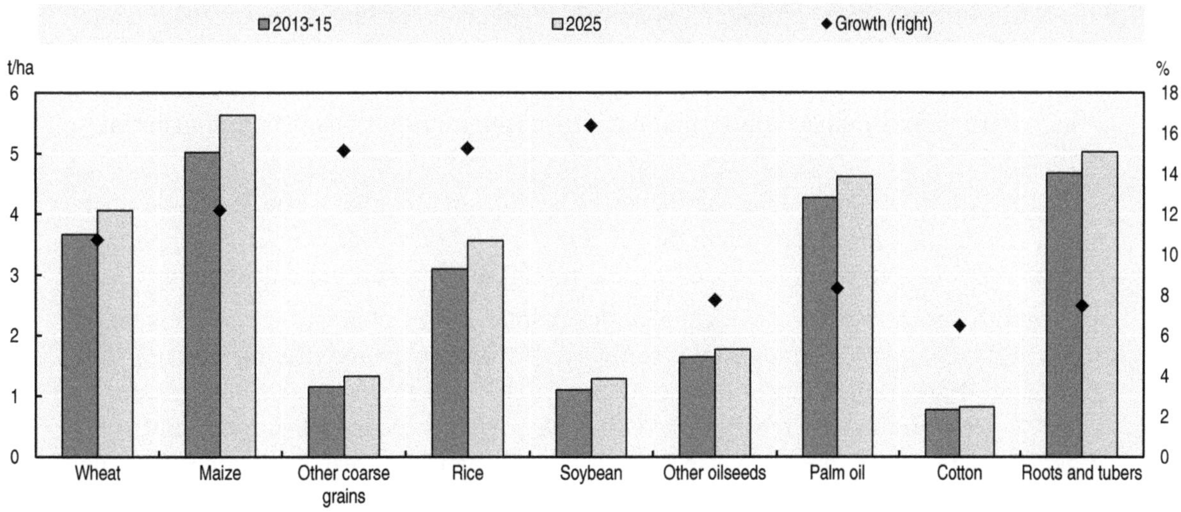

Source: OECD/FAO (2016), "OECD-FAO Agricultural Outlook", *OECD Agriculture statistics* (database), *http://dx.doi.org/10.1787/agr-data-en.*

StatLink http://dx.doi.org/10.1787/888933381219

the region. Neither in India nor any other country of the region do farmers have sufficient incentives to expand their area planted to wheat; all additional output is due to expected yield improvements.

Production of soybeans in South and East Asia will increase by about 30% from its small base. More emphasis is expected in China, India and Indonesia. Still, the region, first and foremost China, will continue importing and crushing vast and increasing amounts of soybeans. In addition to being the leading importer, China is also one of the largest producers of other oilseeds (mainly rapeseed and groundnuts), but output is not projected to expand significantly. Besides producing protein meal and vegetable oils from oilseeds, South and East Asian countries also lead world production of palm oil. As a result of the recent slow-down in Chinese cotton production, India has become the world's largest producer; through further area expansion and the application of new technologies it is expected to produce approximately 30% of the world output by 2025. Even though palm oil will be less used as a feedstock in European biodiesel, production is expected to expand faster than oilseed-based oils mostly to meet domestic mandates.

Satisfying the fast growing domestic demand for meat, dairy and fish remains a challenge for the livestock sector in the region. Meat production will expand by 1.8 Mt annually by 2025, a 17% increase over the base period. Pork and poultry will account for the bulk of the expansion. China continues to be the region's most important livestock producer, especially pork. South and East Asian countries are projected to continue dominating overall aquaculture production, with China, India, Indonesia and Viet Nam to account for the majority of growth over the next decade.

Rapid urbanisation and income growth induced demand in South and East Asia underpins the development of the dairy sector. Milk yield is set to increase by 20% to 2025. Due to a 47% expansion, India will become the world's largest milk producer by 2020. Production of fluid milk in the region strongly outpaces the dairy processing sector since an increasing share of the milk is produced for fresh consumption.

The Americas: Export-oriented and fast growing

North America, and Latin America and the Caribbean combined currently dominate global oilseeds production, with a share of almost 90%, and have significant production shares of around 30% for cereals, meats and dairy. Cereals are more concentrated in the North, particularly maize, while the South focusses more on oilseeds and especially soybeans.

Crop output in North America increases by 10%, led by maize and soybeans. Rice production in the United States is in the process of recovery; wheat remains an important crop as well, but area continues to shrink and yield improvements are only marginal.

Latin America remains the most significant source of agricultural area expansion in the world, with total crop area increasing by 24% and soybeans driving most of this expansion (Figure 1.8). Brazil will become the single most important soybean producer by 2025, with production reaching 135 Mt. The soybean crush will provide not only vegetable oil for export, but also protein meals for an expanding livestock sector. Sugarcane and cotton continue to be a source of growth for Brazilian agriculture through improved yields and expanded area. Prospects for aquaculture are particularly good with an expected growth of 40% between the base period and 2025.

Figure 1.8. **Crop area in Latin America and Caribbean**

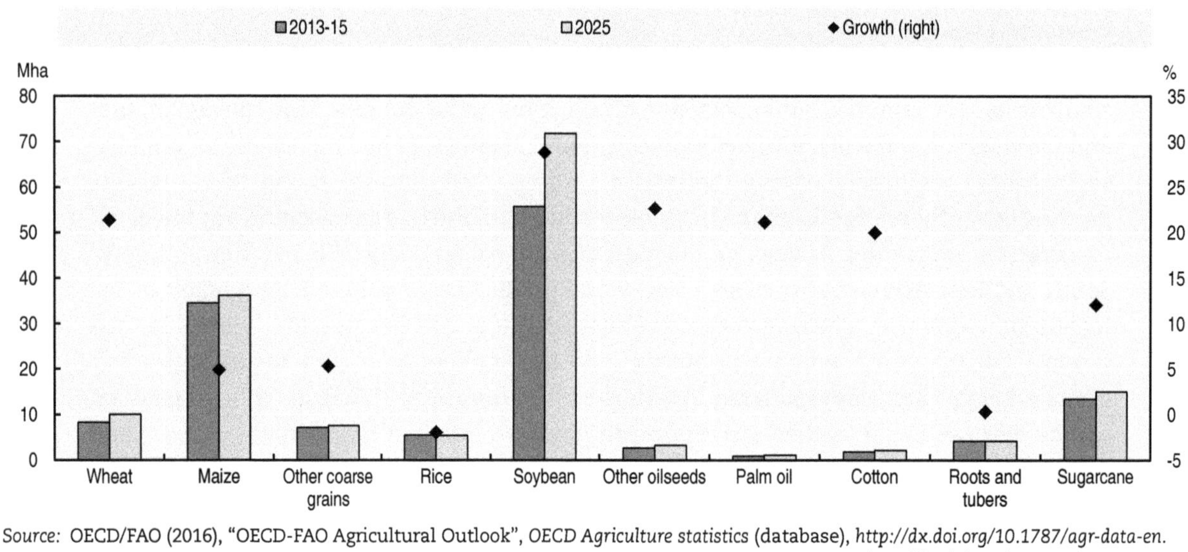

Source: OECD/FAO (2016), "OECD-FAO Agricultural Outlook", *OECD Agriculture statistics* (database), *http://dx.doi.org/10.1787/agr-data-en.*
StatLink *http://dx.doi.org/10.1787/888933381227*

The United States and Brazil remain the two largest ethanol producers. Their growth prospects differ, however, as Brazil is projected to increase its output by 25% during the projection period mainly driven by domestic demand, while the US production is expected to decrease due to weaker domestic and international demand.

Sub-Saharan Africa: Vast potential, but severe constraints

Chapter 2 of this *Outlook* provides an in-depth analysis of the agricultural sector and disaggregated projections for Sub-Saharan Africa. Whereas past production growth has been underpinned by area expansion, over the next ten years area growth is projected to slow, with an increasing share of production growth attributable to improved productivity.

Overall, total agricultural production is projected to expand by 2.6% p.a. in the coming decade. In many instances, production fails to keep pace with domestic demand, resulting in rising import dependence for many primary food products. At the same time, many countries are competitive producers and consistent exporters of fruit and beverage crops, contributing to foreign currency reserves.

Eastern Europe and Central Asia: A global player in cereals

Eastern Europe and Central Asia experienced a rapid expansion of its agricultural output during the previous decade, with crop production increasing by 42%. The region produced 10% of global cereals in 2013-15, mostly wheat. However, the variable weather conditions in the region and ensuing production instability induces uncertainty into global cereals markets.

Over the projection period, total agricultural production will expand by 13%. Natural conditions favour the concentration on cereals, followed by sunflower seed. The rapid area expansion observed during the last decade will not be sustained. Projected shifts in relative profitability will result in crop substitutions. While production gains in sugar crops are based mainly on area expansion, growth in cereals and oilseed is due to increased yields. In general, yields remain below global averages. Soybeans, a relatively new crop for this region, are expected to show the highest yield increase.

Similarly to the crops sector, the livestock sector will further expand in the projection period, albeit at a much lower rate. Total meat production will increase by 3 Mt, with poultry accounting for about one half. Improved milk yields will support the development of dairy processing.

Fuel ethanol and biodiesel will remain marginal in the region as the countries are either oil and gas producers or commodity exporters without incentive policies for biofuel producers or blending mandates for consumers.

Western Europe: Stable production structures

Currently, the industrialised countries in Western Europe hold significant shares of global dairy production (36%), biofuels (30%), meat (15%) and cereals (13%). After years of strong expansion in response to high prices, the projected slowing in crop production results from flat or decreasing biofuel feedstock demand, stagnating domestic feed and food demand and stiffer competition in the world market for cereals, especially from Eastern Europe and Central Asia.

Total harvested area will be reduced by 3% in 2025. Yields are the highest in the world for most crops with only small margins for improvement (4% on average). By 2025, the production in Western Europe is projected to further concentrate on cereals. Individual crop projections are mixed; maize and sugar beet production will expand, while harvests of other coarse grains, other oilseeds, and roots and tubers (mainly potatoes) are expected to decline.

The meat sector is projected to grow at twice the rate of crops, which will result in 1.7 Mt of additional meat production in 2025. Poultry and pork account for most of the gains. Intensification and restructuring of the dairy sector in the European Union will result in an overall expansion of milk production. The strong supply of raw milk backs a significant enlargement of the dairy processing sector. Further growth of the European dairy and meat industry will be determined by its competitiveness in international markets as most of the additional output will be exported.

North Africa and the Middle East: Difficult conditions cause volatility

Unfavourable natural conditions and unstable political situations in many countries in North Africa and the Middle East constrain agricultural production. Owing to a lack of water availability or irrigation the region's share of global production is very low. Wheat is the predominant crop accounting for almost 60% of the harvested area. In countries like Iran and Morocco, local authorities strive to increase self-sufficiency using various policies aimed at boosting wheat production. The progress comes at a cost, as limited resources need to be rationed and the gains in wheat production imply stagnating production of other cereals such as maize and other coarse grains. Sugar crops production also expands in the region, given the importance of sugar as a source of calories. Positive trends are also expected for rice, roots and tubers, but they remain niche products.

The meat and fish sectors will not maintain the growth rate of the previous decade and expand production by only about 1% per year, based overwhelmingly on poultry and sheep. The dairy sector will remain very traditional with low inputs and low yields, utilising cows, camels and goats. Most of the milk is consumed in the form of fresh dairy products, and production of cheese, butter and milk powders remains limited.

Biofuel production is not relevant in the region, because it is oil rich and feedstock crops would have to be imported.

Oceania: Successful niche producer

Oceania's prominence in global agriculture is not the result of its actual production share, but based on its strong per-capita production which generates a high exportable surplus. Even as the preeminent dairy products exporter, its production accounts for only 9% globally; even less for meat (2%) and cereals (2%), its other important export commodities. Oceania is expected to recover from a period of declining production, expanding output by 11% over the projection period. The gains will come mainly from sugarcane and cereals and are mainly based on yield improvements.

Based on a slower growing global import demand for meat, the livestock sector in Oceania is slowing as well. Poultry dominates the expansion but is almost entirely focused on the domestic market; while beef and veal production is projected to remain flat. Milk production will continue to be driven by expanding dairy cow herds, whereas yield developments are slower due to the predominantly pasture-based production system. Capture fisheries will continue to be predominant in Oceania, with a share of 83% of total fish production in 2025.

Trade

Agricultural trade to increase, but at slower rates than in the past

Along with global supply and demand, trade is expected to expand less over the next ten years than in the previous decade (Figure 1.9). The declines are particularly acute for cereals and dairy products, but also apparent for meats (with the exception of sheep meat) and fish. Among non-food products, trade in ethanol and biodiesel is expected to contract, while trade in cotton is projected to recover following sharp declines between 2005 and 2008. The slower growth rates for most food products need to be placed in context, as the trade levels are higher in the base period compared to the previous decade. For example, world meat trade in 2025 is projected to be 22% higher than in the base period and the additional volume of meat traded will be roughly the same in the next decade as in the previous one.

Figure 1.9. **Growth in trade by commodity**

Annual percentage growth in volume terms

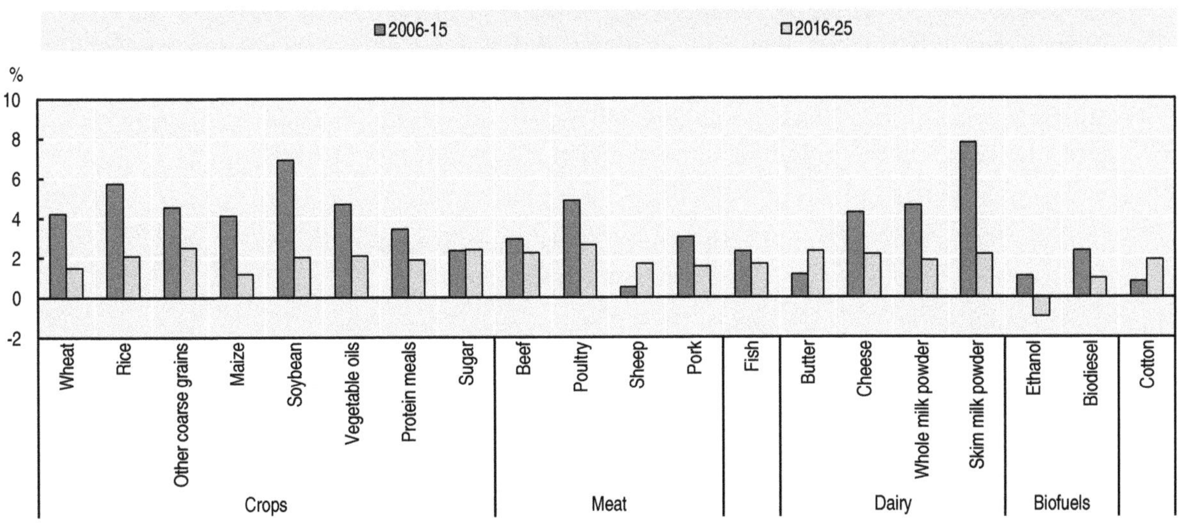

Source: OECD/FAO (2016), "OECD-FAO Agricultural Outlook", *OECD Agriculture statistics* (database), *http://dx.doi.org/10.1787/agr-data-en.*

StatLink http://dx.doi.org/10.1787/888933381231

The main reason for this slowdown is lower growth rates in emerging economies, which have relatively high income elasticities of demand for most food commodities. The importance of China as a major importer of several commodities means that the slowdown in China's growth will have a particularly significant impact (Box 1.5).

A further reason for the slowdown is the adoption of more protectionist policies in some of the larger importing countries. Whereas agricultural trade protection has been declining in most OECD countries, several emerging economies (including China, India and Indonesia) have pursued self-sufficiency objectives and associated import protection. To some extent, the tendency for increased protection may be offset by new regional and plurilateral trade agreements, although agreements may also divert trade away from non-members. Only ratified trade agreements are included in the projections. Hence specific provisions of the WTO Nairobi agreement are included, but the Trans-Pacific Partnership agreement is not. An overview of the Nairobi package is provided in Box 1.4. China's decision in March 2016 to end the stockpiling of maize and release additional maize on the market in the coming years has also been incorporated in the baseline.

The recent appreciation of the US dollar has made US exports less competitive on the world market and reduced export flows from countries whose currencies are linked to the US dollar. On the other hand, the US dollar appreciation can create export opportunities for countries with currencies that depreciate relative to the US dollar. Brazil and Argentina, for example, are projected to expand their shares of the global meat markets, while Australia and New Zealand will maintain their high market share in dairy markets.

The slowdown in trade will not cause any major changes to the share of agricultural production that is traded. Figure 1.10 compares the share of production that was exported during the baseline with the projections for 2025. The ranking of the commodities is not projected to change considerably during the next decade. Whole milk powder (WMP) and skimmed milk powder (SMP) will remain the most traded agricultural commodities and fresh dairy products (not depicted in the figure) will continue to be the least traded. The very low trade in fresh dairy products, with less than 1% of production traded, is directly related to the difficulties in transporting and storing fresh products. Vegetable oils and soybean are also highly traded, with over 40% of their production entering international markets. About 31% of total fishery production is expected to be traded in 2025. Among the different types of meat, beef and poultry will remain the most traded and are projected to account for 80% of the additional meat traded in 2025 relative to the base period.

Figure 1.10. **Share of production traded**

Share of exports in total production

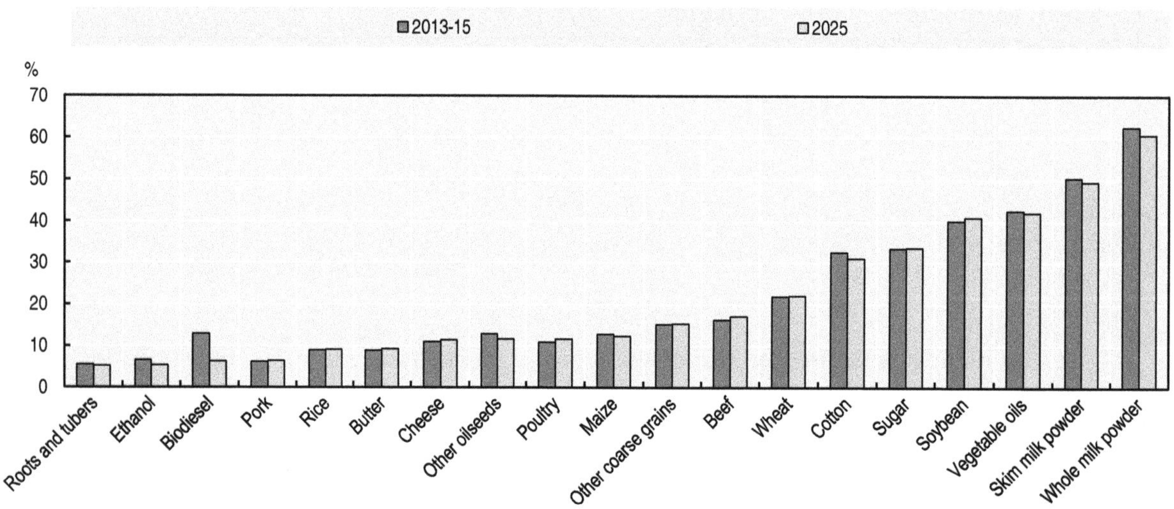

Source: OECD/FAO (2016), "OECD-FAO Agricultural Outlook", *OECD Agriculture statistics* (database), *http://dx.doi.org/10.1787/agr-data-en.*

StatLink *http://dx.doi.org/10.1787/888933381249*

Box 1.4. **What does the Nairobi package imply for agriculture?**

On 19 December 2015 the WTO Tenth Ministerial Conference in Nairobi, Kenya agreed on a package of Ministerial Decisions, a number of which are relevant to agriculture. The "Nairobi Package" includes a commitment to abolish export subsidies for farm exports, in addition to other agriculture-relevant decisions concerning public stockholding for food security purposes; a special safeguard mechanism (SSM) for developing countries; measures related to cotton; and preferential rules of origin.

Export Competition: A key feature of the Nairobi Package is a Ministerial Decision on **Export Competition**, under which developed countries have pledged to eliminate subsidies for farm exports, with the exception of scheduled export subsidies for dairy and processed products and pork. The latter have been given more time and have been agreed to be phased out by the end of 2020. Developing countries have until the end of 2018 to phase out export subsidies, but will be able to continue to cover marketing and transport costs for agriculture exports until the end of 2023. The poorest and food-importing countries will be granted until the end of 2030 to meet their commitments.

Box 1.4. **What does the Nairobi package imply for agriculture?** *(cont.)*

In addition to the above, the decision contains restrictions, or "disciplines", to prevent the use of other export policies as subsidies. These disciplines include limitations on financing support for agriculture exporters, such as export credits, export credit guarantees or insurance programmes; rules for agricultural exporting state enterprises; and disciplines to ensure that international food aid does not adversely impact domestic markets.

Public Stockholding for Food Security: The decision, on **Public Stockholding for Food Security Purposes** reaffirms the commitment of WTO members to negotiate and make all concerted efforts to agree and adopt a "permanent solution" to this issue, which had been at the centre of discussions at the Bali Ministerial in 2013.

Cotton: The **Cotton** decision calls on developed countries – and developing countries that declare themselves able to do so – to grant listed "cotton-related" exports from LDCs duty-free and quota-free access from 1 January 2016 onwards, to the extent provided for in their respective preferential trade agreements in favour of LDCs. Developed countries are also required to end cotton export subsidies immediately, while developing countries must do so by 1 January 2017. The decision also acknowledges reforms made by certain countries to their domestic cotton policies which may contribute to the reduction of domestic subsidies, while emphasising that further efforts need to be made.

Special Safeguard Mechanism (SSM): Countries agreed to maintain the right of developing countries to have recourse to SSM based on import quantity and price triggers with precise arrangements to be further defined as envisaged under paragraph 7 of the Hong Kong Ministerial Declaration. Negotiations on a SSM will be pursued in the WTO Committee.

Other agriculture-relevant decisions at the WTO Tenth Ministerial Conference in Nairobi included a decision on preferential rules of origin for least developed countries. The decision states that when Members apply a processing criterion for agricultural goods they shall, to the extent provided for in their preference programme, allow the transformation of raw agricultural products into processed products to confer origin. Members are also asked to consider extending preferential treatment to products containing non-LDC originating materials of up to 75% of the final value of the product.

Source: World Trade Organization (2015).

Agricultural exports to remain concentrated among a few key suppliers

Traditionally, agricultural exports are concentrated in those few countries that possess the natural endowments necessary for production and have the infrastructure in place to produce and export at competitive prices. During the next decade, that concentration will remain, but there will also be some commodity-specific shifts.

Figure 1.11 illustrates for selected commodities the export shares of each of the top five exporters in 2025. It also indicates the cumulative export share of the top five exporters during the base period (2013-15). In 2025, at least 70% of total exports will originate from only five countries for each commodity. The highest concentration of exports in 2025 is projected to remain in soybean trade, where the top five exporters account for almost 95% of total exports. For most products the cumulative shares of the five biggest exporters are similar to those in the base period, with some slight declines (e.g. wheat and cotton) and some increases (e.g. cheese, sugar and SMP).

The importance of a few exporters for the supply of numerous commodities implies potentially significant market impacts if exports are interrupted, either as a result of production shocks or policy changes. Those risks are particularly acute in the case where just one or two countries hold a significant share of exports, and other countries may struggle to replace a shortfall, at least in the short-term. In 2025, just one country is projected to account for more than 40% of world exports of roots and tubers (Thailand), WMP (New Zealand), butter (New Zealand), other oil seeds (Canada), sheep meat (Australia), and sugar and soybeans (Brazil).

Figure 1.11. **Export shares of top 5 exporters in 2025, by commodity**

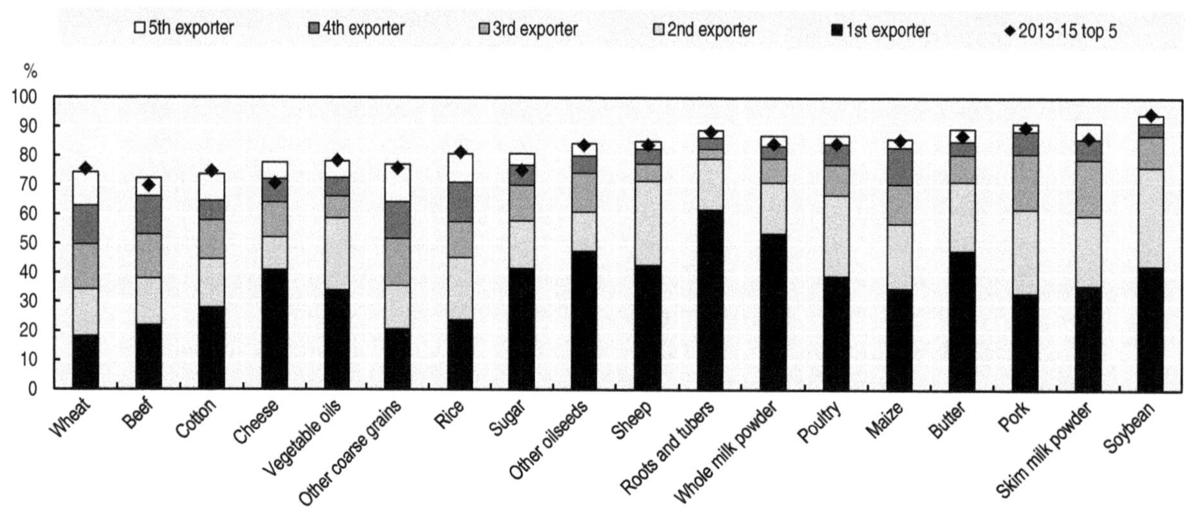

Source: OECD/FAO (2016), "OECD-FAO Agricultural Outlook", *OECD Agriculture statistics* (database), *http://dx.doi.org/10.1787/agr-data-en.*

StatLink http://dx.doi.org/10.1787/888933381258

The composition and ranking of the top three exporters remains unchanged for most commodities between the baseline and 2025, with a few exceptions. The United States will remain the main exporter of maize, but will lose some market share to Brazil. The top three exporters of rice during the base period – India, Thailand and Viet Nam – were responsible for over 65% of total exports. By 2025, Viet Nam and India will have swapped places, making Viet Nam the largest exporter, and the export share of the top three exporters will become less than 60%. This is a result of the emergence of Cambodia and Myanmar as major rice exporters. Brazil is projected to replace the United States as the principal exporter of soybeans and India as the main beef exporter. One of the reasons behind these shifts is the ongoing depreciation of the Brazilian real which makes its exports more competitive.

Agricultural imports to be more dispersed, but with China a key market for several commodities

With consumption growth projected to outpace production growth in many countries, imports will continue to be more dispersed among countries than exports (Figure 1.12). For some commodities, however, most notably soybeans, and roots and tubers, a relatively high share of import demand comes from just a few countries. China is a major importer of several commodities, and accounts for a large share of the markets for soybeans and

Figure 1.12. **Import shares of top 5 importers in 2025, by commodity**

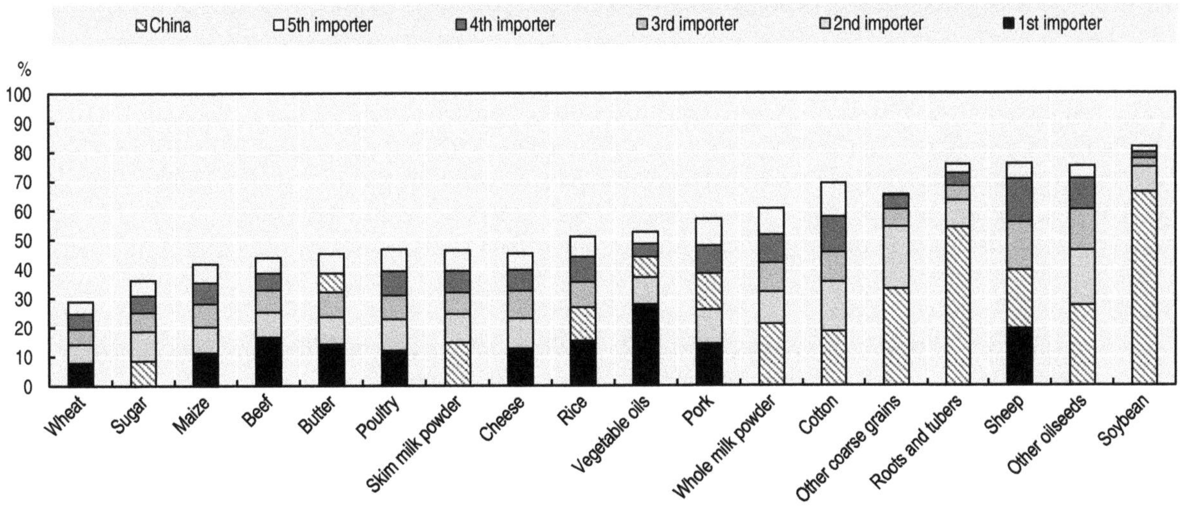

Note: Shading for China is super-imposed depending upon its position among leading importers.
Source: OECD/FAO (2016), "OECD-FAO Agricultural Outlook", *OECD Agriculture statistics* (database), *http://dx.doi.org/10.1787/agr-data-en.*
StatLink *http://dx.doi.org/10.1787/888933381265*

other oilseeds, roots and tubers, other coarse grains, cotton and milk powders. The influence of China's economic growth and of its grain policies is further explored in Box 1.5.

The high concentration of imports for oilseeds and cotton is tied to large scale processing of these commodities in a few countries. Soybean imports to China are projected to account for more than 65% of world imports by 2025, a 105 Mt increase compared to the baseline. The largest demand for cotton imports will also come from China in 2025, even though Bangladesh is projected to be a close second, followed by Viet Nam and Indonesia. These latter three countries will increase their cotton imports considerably compared to the baseline.

China is the largest producer of sheep meat and pig meat in the world, yet also imports large amounts of both types of meat. Over the next decade, China is projected to further increase its domestic production of meat as well as it imports. In the case of beef and sheep, which are the meat types that require the most feed, China's imports will even exceed its domestic production. China will remain the largest importer of SMP and WMP in 2025, however its import shares of WMP are projected to decrease from 25% in 2013-15 to 21% in 2025. Viet Nam, Algeria and Nigeria are projected to emerge as major importers of WMP.

Box 1.5. **The implications of economic developments and cereal policy changes in China**

The importance of China for world agriculture has been stressed in previous *Outlook* editions, in particular the 2013 report, which had a special focus on prospects and challenges for the country's agricultural sector. In the current *Outlook*, China's economic growth rate is expected to slow to an average of 6% p.a. between 2016 and 2025, compared with an average of 9% over the previous decade. Under the baseline, cereals imports are projected to grow by 7% per year on average, which is slower than the growth rate between 2006 and 2015. This slowdown is a major factor behind the projections for slower agricultural trade growth and easing international prices.

Box 1.5. **The implications of economic developments and cereal policy changes in China** *(cont.)*

If GDP growth slows further to 5% per year that will have a significant implication for several markets. A scenario analysis using the Aglink-Cosimo model finds that such a growth reduction would result in prices for most commodities being between 1% and 4% lower in 2025 compared to the baseline projections. Under this scenario, the greatest declines would occur for roots and tubers (-6.3%), oilseeds other than soybeans (-3.8%) and sheep meat (-3.7%). Price declines would be smaller for most meats (less than 3% for all except sheep meat) and for dairy (less than 2%). Across commodities, China's share of world imports would slip by 1-2%, with a relatively large decline in market share for pork meat (where the share would fall to 8%, versus 12% under the baseline).

Figure 1.13. **World price changes in 2025 under a lower China GDP growth scenario**

Percentage difference compared to the baseline

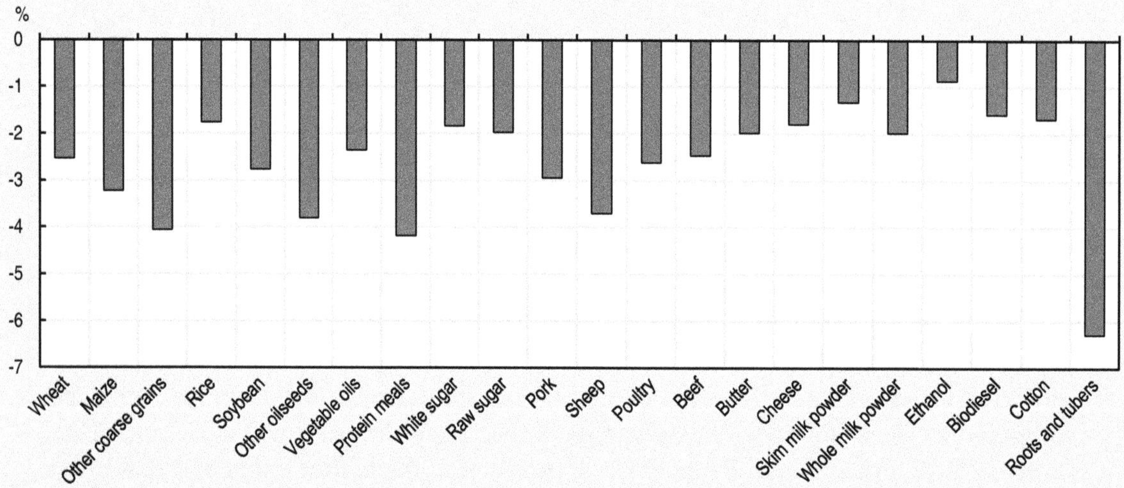

Source: OECD and FAO Secretariats.

StatLink http://dx.doi.org/10.1787/888933381271

Among cereals, the strongest world price impact would be found in the other coarse grains markets, with prices 4.2% lower by 2025. The second largest impact would be found in maize markets (-3%). Although Chinese imports are much lower than for barley and sorghum, the reductions would be stronger in absolute terms (1.8 Mt vs 1.2 Mt). This is due to Chinese maize production reacting much less elastically to price changes than import supply. International prices for wheat and rice would be reduced by 2.5 and 1.9% respectively.

Recently announced changes to China's cereals policies could have a significant impact. Since 1996, China's share of world cereal markets has been constrained by a policy of maintaining a 95% self-sufficiency ratio for key grains, including wheat, maize and rice. Various policies have been put in place to achieve this objective, including:

- Minimum prices for rice and wheat, ad hoc interventions for maize, direct payments to support grain production, transfer payments to major grain-producing counties, comprehensive subsidisation of agricultural inputs as well as buffer stock norms and a few planting directives. Although price support also exists for cotton and oilseeds, persistently higher returns for grains have led to land being allocated to grain production, especially maize.
- Tariff rate quotas (TRQs) for rice, wheat and maize, which were introduced at the time of China's WTO accession in 2001. TRQs amount to roughly 5% of domestic demand and are subject to much lower import tariffs compared to out of quota imports. These TRQs have limited imports of the three main grains, but led to increasing imports of barley and sorghum as well as dried distillers' grains, and roots and tubers (cassava) for feed, where no import quotas exist.

Box 1.5. **The implications of economic developments and cereal policy changes in China** *(cont.)*

Under these domestic and trade policies, support to the grains sector has increased over the years, with domestic prices being roughly one-third higher than those prevailing on world markets. This has led to the accumulation of substantial surpluses, with maize stocks rising from an estimated 45 Mt in 2005 to over 100 Mt in 2015. By 2013-15, stock-to-use ratios had reached 40% for wheat, 45% for maize and above 60% for rice.

China has required increasing imports of grains in order to meet rising demand for animal feed. Until 2009, these imports were no more than 2 Mt and came from a combination of maize, distillers dry grains (DDGs), sorghum, but mainly barley. Overall these imports accounted for about 2% of their total feed use. By 2014 those imports had risen to 30 Mt, accounting for nearly 20% of total feed use. Imports of DDGs, sorghum and barley are almost all for feed use and account for world markets import shares of 80%, 75% and 15% respectively. By contrast, maize imports represent a much smaller share – about 4% – of the world market. Roots and tubers (cassava) is another important feed ingredient imported by China. China's share in world imports has increased between 2000 and 2005 from 10% to over 50% of world imports and total Chinese imports reach 8.8 Mt dry matter in 2014.

With stocks rising and increasing food demand, China has decided to align domestic maize prices more closely with world prices, and maize farmers will receive a deficiency payment equal to a different between the market price and a target price from 2016 onwards. The abolition of minimum prices and the unavoidable release of stocks will lower domestic prices. If the stocks-to-use ratio were to fall to a more sustainable 30% (implying a total of 66 Mt) then about 35 Mt would need to be released. The release of stocks would tend to lower domestic prices, but some of the effect will be offset by increased domestic quantities demanded at lower prices. If stocks are released gradually (say at 5 Mt per year) that would add 4% to annual trade (currently 130 Mt) and 0.5% to world supplies (which run at 1 000 Mt). The substitution of maize for barley, sorghum and DDGs would potentially result in much bigger effects on these markets.

Over the longer term, China faces resource and environmental constraints to increasing its maize production, so imports would be expected to increase. The current *Outlook* assumes that China will reform its policies in a way that addresses its domestic objective of maintaining a high self-sufficiency ratio yet without severely disrupting international maize markets. However, the timing and scale of stock release is a major uncertainty underlying the projections.

Importance of trade for food security

As patterns of consumption and production continue to evolve, global trade in agricultural products is expected to continue to increase over the coming decades. Trade will influence the extent and nature of food security across all regions of the globe. In 2025, about one quarter of the world's population obtain at least 25% of their cereal food use from imports.

Resource constrained countries in the Middle East and Northern Africa (MENA) and a number of countries in Sub-Saharan Africa are especially dependent on imports of basic and high value food commodities. The MENA region currently imports more than half of its main staple food, wheat; 70% of sugar; and 80% of vegetable oils. The dependency on wheat imports is expected to ease slightly in the coming decade, while it will increase for other products. In Sub-Saharan Africa, the need for imports is especially high for vegetable oils (50%), poultry meat (36%) and sugar (23%), all with an increasing trend.

As food imports increase further, many countries become increasingly concerned about the reliability of global markets as a source of affordable food. Opening to trade increases food availability in importing countries and exerts downward pressure on consumer prices. On balance it reduces risks, as global markets tend to be less volatile than domestic markets, although greater reliance on international markets can leave countries vulnerable to the actions of trading partners and to external market shocks.

Prices

Nominal prices to increase by 2025, but remain below recent peak levels

The *Outlook* uses prices at main markets (e.g. US gulf ports, Bangkok) of each commodity as international reference prices. Historic observations are used to describe previous developments while projected values reflect future market trends. Near term price projections are still influenced by the effects of recent market events (e.g. droughts, policy changes), whereas in the outer years of the projection period, they are driven by fundamental supply and demand conditions only.

Figure 1.14 shows for selected commodities the projected evolution of nominal prices. Over the short-term, a combination of rebuilt global stocks and sluggish demand will keep prices for rice, wheat and other coarse grains under pressure. By contrast, maize prices are not expected to fall any further in 2016, after a sharp drop in 2015. Over the medium-term, prices for all cereals are projected to follow a similar trend of modest nominal increases, with more significant gains in other coarse grains, which is mostly a result of the high demand for animal feed in China and limited production expansion possibilities in the main production regions.

Sugar prices remain stable, with global production projected to meet the growing demand in developing countries. The white sugar premium is projected to increase at the start of the projection period following growing import demand. However this situation is expected to change in 2017 due to the abolishment of the sugar quota in the European Union which will lead to lower raw sugar imports and higher white sugar exports. Towards the end of the projection period, the premium is projected to increase again as producers switch to exporting more raw sugar than white sugar.

Prices for protein meals grow at a faster rate than the price of vegetable oil. Whereas vegetable oil consumption is driven mainly by food demand increases in developing countries, the growth in protein meal demand is stronger due to increased non-ruminant and milk production and a greater incorporation of protein in feed rations in developing countries. Additionally, the sustained expansion of palm oil production depresses vegetable oil prices. As a result, soybean prices also rise at a faster rate than the prices of other oilseeds, reflecting their greater protein meal content.

Meat prices decline in the short-term as producers respond to lower feed grain prices. Owing to feed intensive production practices, pork and poultry producers gain a greater benefit from lower feed grain prices, while the shorter production cycle of poultry in particular allows a more rapid response to price signals. Its efficiency in feed conversion and rate of efficiency gains allows poultry to remain the cheapest source of protein, despite firm demand growth. Within the beef sector, which exhibits the longest production cycle, prices trend downwards until 2019 before recovering modestly to 2025. Cheese prices maintain a significant premium over other dairy products, owing to strong demand in both developing and developed economies.

Fish prices are projected to decline in the short-term, then stabilise. The tightening constraints of capture fisheries compared to the vastly expanding aquaculture are

Figure 1.14. **Agricultural prices in nominal terms**

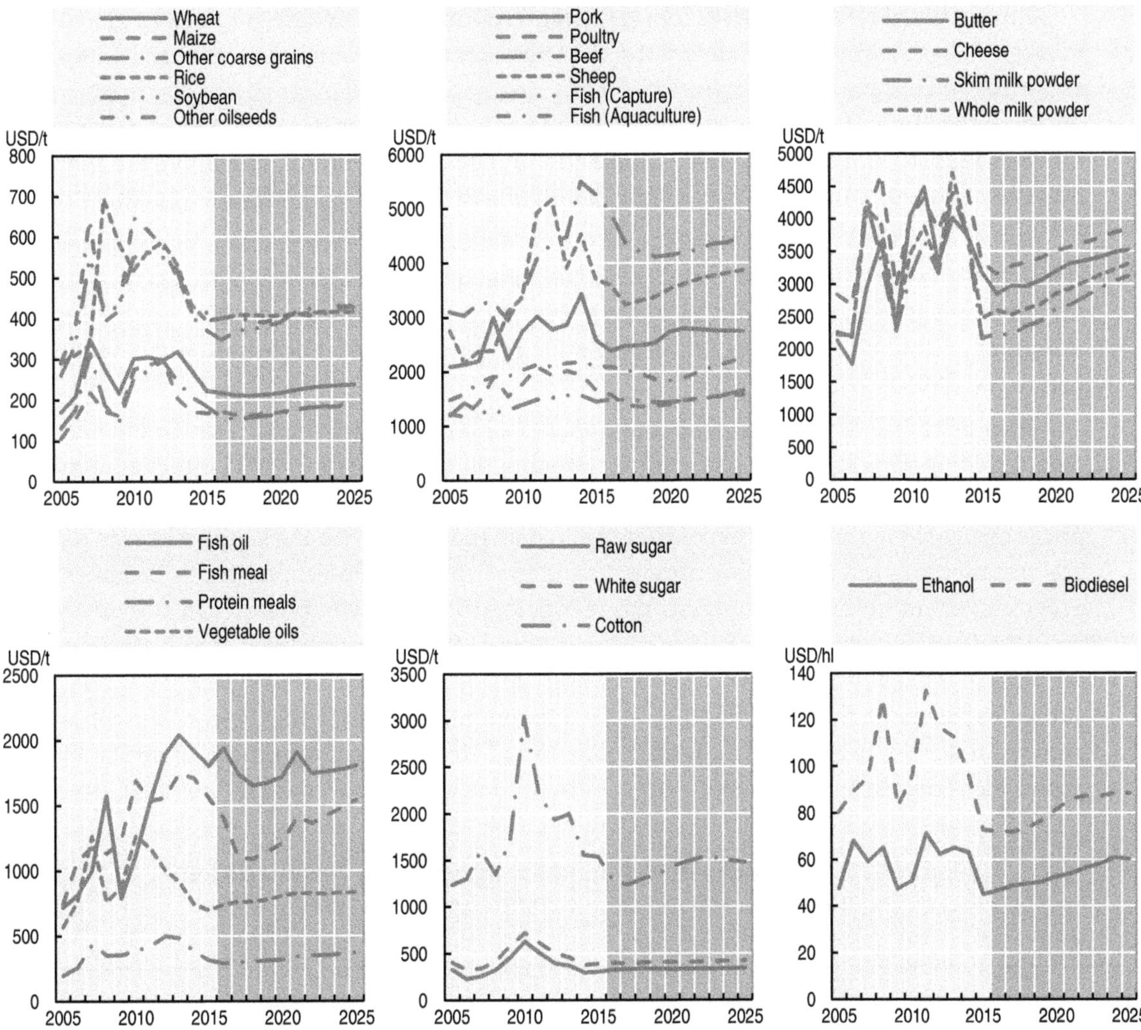

Source: OECD/FAO (2016), "OECD-FAO Agricultural Outlook", *OECD Agriculture statistics* (database), *http://dx.doi.org/10.1787/agr-data-en.*

StatLink http://dx.doi.org/10.1787/888933381284

reflected in the projected prices. The average price for fish caught in the wild is expected to be stronger than that of farmed fish.

The world ethanol price is projected to increase at a faster rate than most other agricultural commodities, reflecting the faster recovery of the crude oil price. However, upward price pressure will be constrained by relatively modest global import demand and strong export potential from the United States and Brazil. Biodiesel prices are closely linked to the vegetable oil prices. Demand for biodiesel is mostly driven by policies, which supports prices of both vegetable oil and biodiesel. However, the end result is that prices do not rise as strongly as for ethanol.

World cotton inventories have reached over 80% of annual consumption. Due to this current over-supply in the market, prices are projected to decrease between 2016 and 2018 and to increase thereafter. These price increases are limited by the increasing competition with man-made fibres and will be strongly influenced by China's future cotton policy.

Modest changes in real prices for most agricultural commodities

International reference prices for agricultural commodities provide guidance to global markets. These signals are transmitted to producers and consumers in the individual countries and influence their market decisions. The transmission of these signals depends on the integration of the domestic markets into the world market. Ultimately, domestic markets react to real domestic prices. Their trend may differ from the global price signal because of real exchange rate movements, and domestic market conditions and policies.

When evaluated in constant US dollars, international reference prices will not divert significantly from their current levels over the next ten years. Globally, both production and demand growth are projected to slow, implying a broadly neutral price path. The overall expectation is that real prices will still remain slightly higher than in the years before the 2007-08 price spike.

Real price trends of individual commodities depend on their particular supply and demand situation. Figure 1.15 shows the average annual change in real international reference prices over the projection period. Globally, the impetus for lower commodity prices comes from a combination of supply and demand factors. On the supply side, yields are assumed to grow at current trends. In the most productive economies, it becomes progressively more difficult to push the technological frontier outwards. On the other hand, in developing countries there is continued scope for yield catch up via more efficient farming practices. Improvements in Asia and Latin America are crucial for the global supply expansion. On the demand side, population growth is slowing, so too is income growth in developing countries, where consumers also have a declining propensity to spend their income gains on food.

Figure 1.15. **Annual price change for agricultural commodities**

Measured in real US dollars (2010) between 2016 and 2025

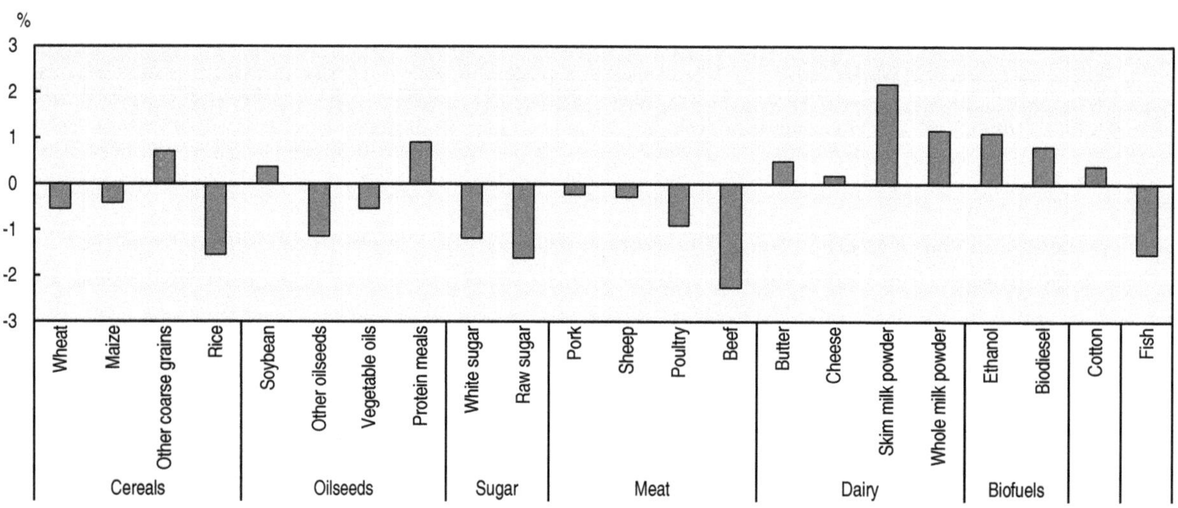

Source: OECD/FAO (2016), "OECD-FAO Agricultural Outlook", *OECD Agriculture statistics* (database), *http://dx.doi.org/10.1787/agr-data-en.*

StatLink *http://dx.doi.org/10.1787/888933381293*

Cereal prices, with the exception of other coarse grains, are projected to decline in real terms. Meat prices will trend moderately downward based on a slower demand growth and fast industrialisation of production. The comparably stronger demand for dairy products and required costly expansion of production capacity push prices for dairy products higher over the next ten years. Rising import demand, particularly from Asia and Africa, supports a recovery in milk powder prices from recent lows. Biofuel prices follow the crude oil and feedstock price trends.

The long-term trend for agricultural prices can be difficult to discern and depends on the historical perspective. For example, prices are projected to be structurally higher than in the decade before the 2007-08 price spike, but not inconsistent with a very long-term trend for declining real prices (Figure 1.16). Over the past 100 years, wheat prices have declined by an average of 1.5% per year in real terms. While the broad tendency over the past century has been for prices to decline, markets have been subject to shocks (wars, political crises and natural disasters) that caused periods of rising prices and high volatility. The magnitude of these shocks has been declining over time due to improvements in the resilience of production and the globalisation of trade. Similar patterns are evident across other commodities.

Figure 1.16. **Long-term price of wheat in real terms**

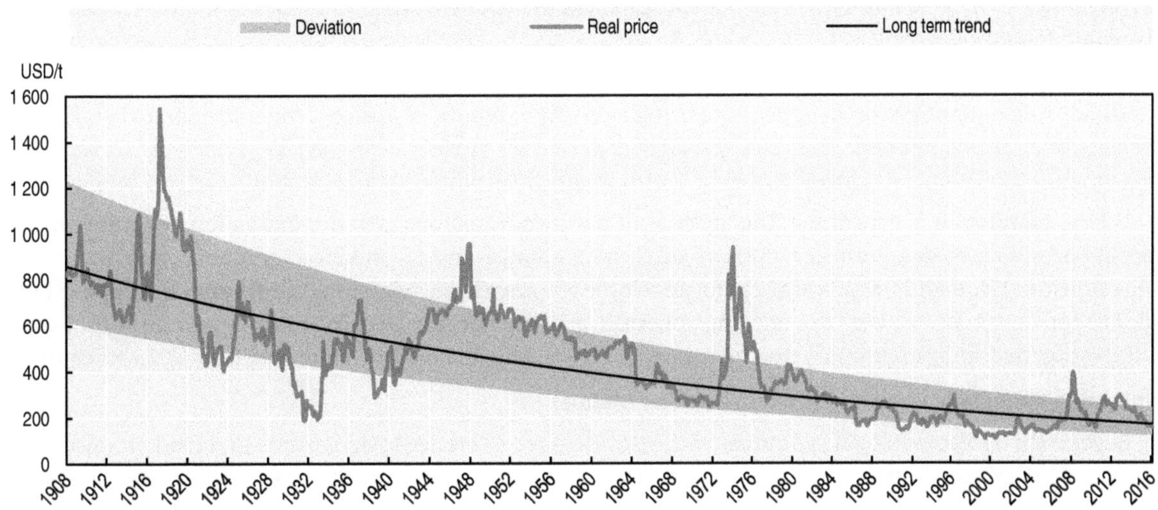

Note: Deviation refers to 1 standard deviation above and below the trend line. Assuming the distribution of error terms is normal then 68% of visualisations should occur within this range.

Source: Monthly "Wheat price received" from USDA quickstats. Deflated using monthly CPI data from *www.bls.gov/data/*.

StatLink http://dx.doi.org/10.1787/888933381300

The baseline projections in the *Outlook* assume a stable market environment. In order to examine the sensitivity of the projections to the variability in macro-economic conditions and yield fluctuations, a complementary uncertainty analysis is performed. This analysis accounts for potential variability in economic growth, exchange rates, yields and oil prices, with future variability of these elements assumed to be the same as in the past. Other sources of variability, such as animal diseases or policy changes, are not considered. Climate change is accounted in the analysis to the extent that it is reflected in yield trends and existing yield variability. This implies that the potential impacts of climate change in terms of shifts in production to new areas, or greater yield variability, are not included in the analysis. The wider potential implications of the recent COP21 agreement are discussed in Box 1.6.

Box 1.6. **What are the implications of COP21 for agriculture?**

At the COP 21 in Paris, agreement was reached on the UNFCCC[1] Paris climate accord in December 2015. The Paris Agreement sets a long-term goal to contain the increase in global average temperatures to well below 2°C above pre-industrial levels and a pledge to "pursue efforts" to limit it to 1.5°C. To reach this goal, parties agreed on the need for global emissions to peak and start declining as soon as possible – recognising that this will take longer for developing countries – and to undertake rapid reductions thereafter in accordance with the best available science.

Agriculture is not directly mentioned within the agreement itself. Nevertheless, both the text and the country-level strategies for emissions reduction, which are outlined in the form of Intended Nationally Determined Contributions (INDCs), recognise the threat which climate change poses to sustainable food production and **offer opportunities for agriculture to be an active part of the solution to climate change.**

Relevance of the Paris Agreement for food and agriculture

Explicit reference is made within the preamble of the agreement to food security and production, which acknowledges "the fundamental priority of safeguarding food security and ending hunger, and the particular vulnerabilities of food production systems to the adverse impacts of climate change". Moreover, Article 2 of the agreement underlines the importance of food production, clearly stating that "This agreement (...) aims to strengthen the global response to climate change (...) in a manner that does not threaten food production".

By giving governments the freedom to decide exactly which emission sources to address, the agreement does not rule out **mitigation** in agriculture. Article 4.1, for example, states governments' aim to "*achieve a balance between anthropogenic emissions by sources and removals by sinks of greenhouse gases in the second half of this century*". Reference is made in Article 5.1 to carbon sinks which should be conserved and enhanced.

Where **adaptation** is concerned, the agreement outlines numerous government actions to strengthen societies' ability to deal with the impacts of climate change and to provide continued and enhanced international support for adaptation to developing countries. These include **financial support** by developed countries, such as the ongoing collective goal to mobilise USD 100 bln per year until 2025 for adaptation and mitigation in developing regions, a figure which should be increased from 2025 onwards.

Intended Nationally Determined Contributions (INDCs)

Beyond the agreement itself, a number of the INDCs make reference to agriculture and food production. Of the 133 INDCs analysed by the Consultative Group of International Agricultural Research (CGIAR) in late November 2015, agricultural adaptation was referred to in 102 (94 of which included at least one adaptation measure), and targets related to agricultural mitigation were included in 103 (84 of which specified at least one mitigation measure).[2] Agricultural water management was included in 83 submissions, while forestry featured in 153 INDCs.

The application of the INDCs will be supported by the **Lima-Paris Action Agenda (LPAA).** The LPAA features five major initiatives concerning agriculture. Initiatives include the 4 per 1000 Initiative: Soils for Food Security and Climate, launched by state and non-state partners, which aims to protect and increase carbon stocks in soils, and the Adaptation for Smallholder Agriculture Program (ASAP), which intends to increase the climate resilience and food security of smallholder farmers.

1. United Nations Framework Convention on Climate Change.
2. CGIAR, Research Program on Climate Change, Agriculture and Food Security, and CCAFS (November 2015), *https://cgspace.cgiar.org/rest/bitstreams/62364/retrieve*.

Results of the uncertainty analysis are illustrated for two commodities in Figure 1.17, which shows the evolution of the nominal prices of maize and pork meat, together with the variation around the baseline. Yield uncertainty is assumed to be constant over time, whereas macro-economic uncertainty accumulates over time. This results in the overall uncertainty tending to be greater towards the end of the projection period. The impact of yield uncertainty and macro-economic uncertainty varies by commodity, with macro-economic uncertainty overall exerting a larger influence on prices than yield variation. For meat and dairy products, the effect of crop yield uncertainty on world market prices is very low compared to macro-economic uncertainty.

Figure 1.17. **Prices in nominal terms for maize and pork**

Including variation derived from stochastic analysis

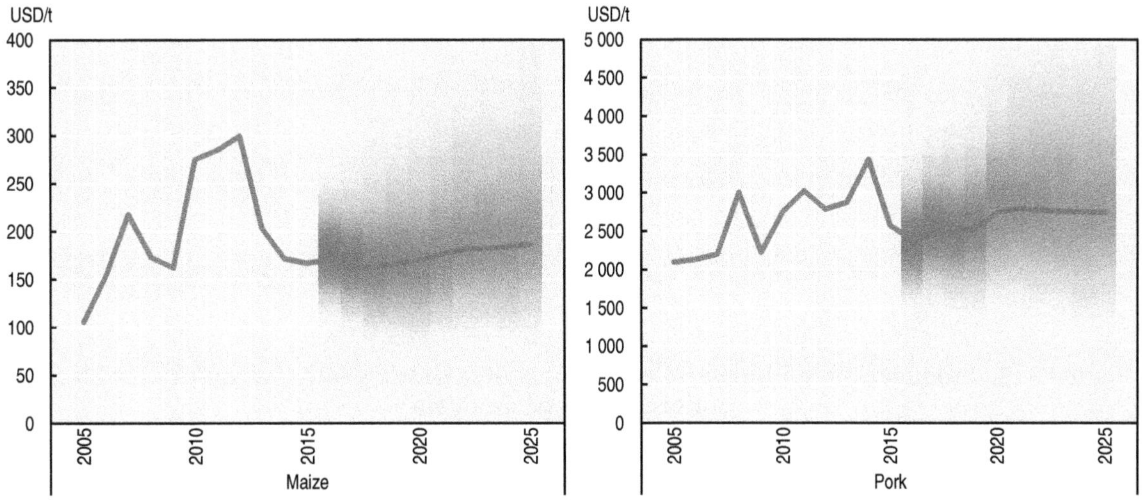

Source: OECD/FAO (2016), "OECD-FAO Agricultural Outlook", *OECD Agriculture statistics* (database), *http://dx.doi.org/10.1787/agr-data-en.*

StatLink http://dx.doi.org/10.1787/888933381310

Food consumer prices likely to exert less volatility than agricultural producer prices

The OECD-FAO Agricultural Outlook projects the evolution, for major agricultural commodities, of prices at the producer and final consumer levels over the medium-term. The representation of consumer prices in the current Outlook has been modified to better reflect the process of food price formation along the food chain and the fact that margins between producer and consumer prices vary across commodities and countries. The modelling of final food consumer prices takes into account the evolution of agricultural prices, the determinants of margins (in the form of labour and transportation costs) and of the ability of consumers to substitute among food products. Overall, margins are projected to increase over the projection period for most countries and commodities, implying that the ratio of the producer price to the consumer price declines. This is because other costs (for fuel and transformation) increase more rapidly than producer prices.

As an illustration, Figure 1.18 presents the historical and the projected evolution of aggregate PPI and CPI measures for a selection of countries and regions.[1] Between 2005 and 2015, PPI growth exceeded CPI growth, except in the European Union. Over the last decade, the variability of PPI measures has been higher than the variability of food CPI measures. Stochastic analysis suggests that this will continue over the next ten years, as illustrated for the case of Mexico (Figure 1.19). Future editions of the *OECD-FAO Agricultural Outlook* will explore the relationships between producer and consumer prices in more detail.

Figure 1.18. **Producer price index and food consumer price index for selected countries**

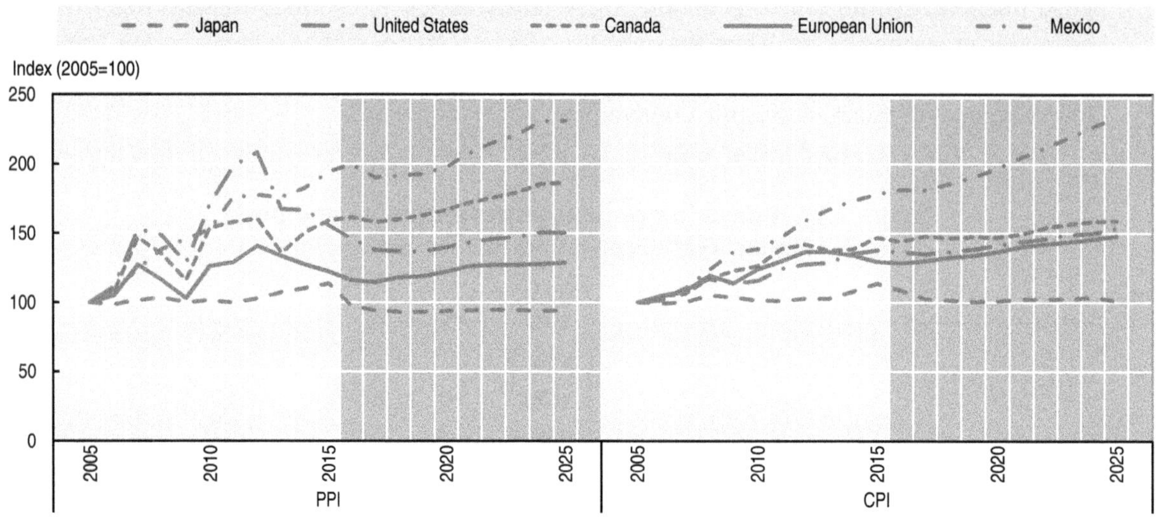

Source: OECD/FAO (2016), "OECD-FAO Agricultural Outlook", *OECD Agriculture statistics* (database), *http://dx.doi.org/10.1787/agr-data-en.*

StatLink *http://dx.doi.org/10.1787/888933381327*

Figure 1.19. **Producer price index and consumer price index yearly variability for Mexico under stochastic analysis**

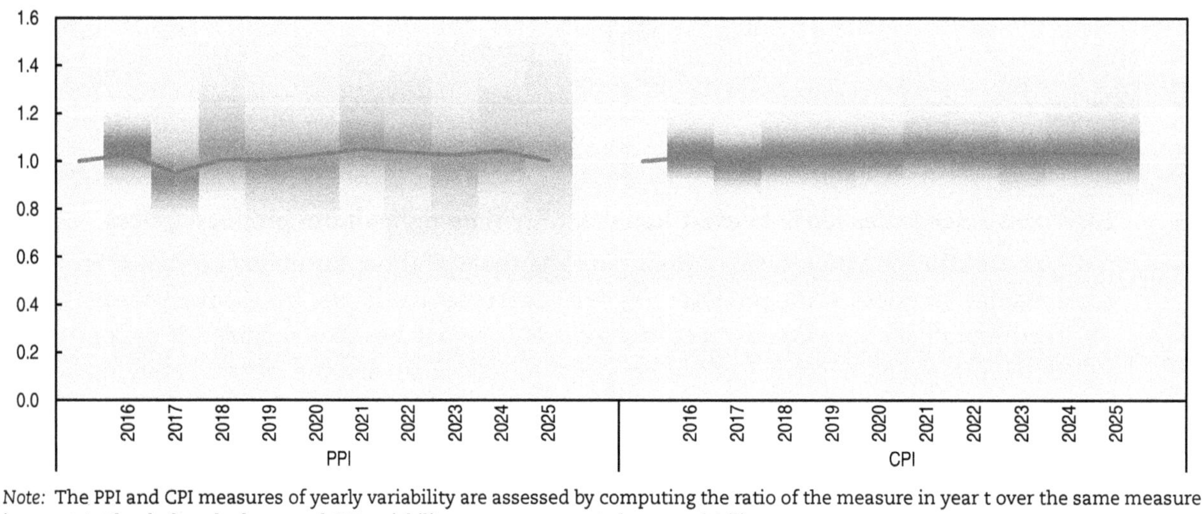

Note: The PPI and CPI measures of yearly variability are assessed by computing the ratio of the measure in year t over the same measure in year t-1. The darker shade around CPI variability measure suggests lower variability.

Source: OECD/FAO (2016), "OECD-FAO Agricultural Outlook", *OECD Agriculture statistics* (database), *http://dx.doi.org/10.1787/agr-data-en.*

StatLink *http://dx.doi.org/10.1787/888933381330*

Conclusion

Over time, the share of raw product value in retail food products has declined. This trend is projected to continue over the next ten years. At the same time, farm prices will remain more volatile than consumer prices due to several factors including weather-dependency (crops) and long production cycles (livestock).

The continued demand for agricultural commodities is projected to be met by efficiency gains in production which will keep real agricultural prices relatively flat. Nevertheless, there will be some relative price changes reflecting changes in the composition of demand and supply constraints. Thus, livestock prices are projected to rise relative to crop prices, and the prices of coarse grains and oilseeds are projected to rise relative to the prices of food staples. Those structural trends are likely to be more apparent in the current context of low prices across all commodity groups.

Note

1. Aggregate measures of Producer Price Indices (PPI) and Consumer Price Indices (CPI) have been computed on the complete *Outlook* database with weights determined on the basis of production and consumptions shares respectively.

Reference

OECD (2015), *Aglink-Cosimo model documentation*, *www.agri-outlook.org/abouttheoutlook/Aglink-Cosimo-model-documentation-2015.pdf*

OECD-FAO Agricultural Outlook 2016-2025

PART I

Chapter 2

Agriculture in Sub-Saharan Africa: Prospects and challenges for the next decade

This chapter reviews the prospects and challenges facing the agricultural sector in Sub-Saharan Africa over the next decade. It reviews sector performance, outlines the current market context, provides detailed quantitative medium term projections for the ten-year period 2016-25, and assesses key risks and uncertainties. The outlook for agriculture in Sub-Saharan Africa is situated in the context of several mega-trends that shape the sector's development. These include rapid population growth, urbanisation and rural diversification, an associated structural transformation from farm to non-farm employment, a growing middle class, and increasing interest (both domestically and globally) in the continent's farmland. The Outlook for agriculture is broadly positive, but could be further enhanced by consistent policies and strategic investments, in particular in rural infrastructure.

The statistical data for Israel are supplied by and under the responsibility of the relevant Israeli authorities. The use of such data by the OECD is without prejudice to the status of the Golan Heights, East Jerusalem and Israeli settlements in the West Bank under the terms of international law.
The position of the United Nations on the question of Jerusalem is contained in General Assembly Resolution 181(II) of 29 November 1947, and subsequent resolutions of the General Assembly and the Security Council concerning this question.

Introduction

The Sub-Saharan Africa[1] (SSA) region accounts for more than 950 million people, approximately 13% of the global population. By 2050, this share is projected to increase to almost 22% or 2.1 billion. Undernourishment has been a long-standing challenge, with uneven progress across the region. Despite being reduced from 33% in 1990-92 to 23% in 2014-16, the percentage of undernourishment remains the highest among developing regions (FAO, IFAD and WFP, 2015). Owing to rapid population growth of 2.7% p.a. over the same period, the absolute number of undernourished people has increased by 44 million to reach 218 million. Slow progress towards food security has been attributed to low productivity of agricultural resources, high population growth rates, political instability and civil strife. However, vast regional differences remain and the success achieved in countries with stable political conditions, economic growth and expanding agricultural sectors suggests that appropriate governance systems, institutional capacities, and macro-economic, structural and sectoral policies can work together to improve food security on a long-lasting and sustainable basis.

The important role of the agricultural sector in contributing to food security is reflected in its prioritisation in the development agenda. The Comprehensive African Agricultural Development Programme (CAADP) is an integral part of the New Partnership for Africa's Development (NEPAD) and the sector's prominence in the region is evident in its contribution to total GDP, which is generally high in the global context. The high contribution of the agricultural sector to GDP also underlines the limited diversification of most African economies. On average, agriculture contributes 15% of total GDP, however it ranges from below 3% in Botswana and South Africa to more than 50% in Chad (Figure 2.1), implying a diverse range of economic structures. Agriculture employs more than half of the total labour force (IMF, 2012) and within the rural population, provides a livelihood for multitudes of small-scale producers. Smallholder farms constitute approximately 80% of all farms in SSA and employ about 175 million people directly (Alliance for a Green Revolution in Africa, 2014). In many of the countries, women comprise at least half of the labour force (FAO, 2015).

Given its role in confronting the challenge of eradicating hunger and improving food security, this chapter considers the historic performance and current state of agriculture in SSA, within the context of the region's political and economic conditions, natural resource situation and demographic structure. It acknowledges the role of policies and megatrends[2] in shaping development of the agricultural sector. Megatrends include demographic change, the rise of the African middle class, growing access to new information and communication technologies, rapid urbanisation and consequent shifts in food demand. This is accompanied by downstream modernisation of food systems, a considerable shift in the labour force from farming to nonfarm jobs, and rising global interest in available African farmland strengthened by the sharp rise in agricultural commodity prices over the past decade.

Figure 2.1. **Agriculture as a share of total GDP in 2014**

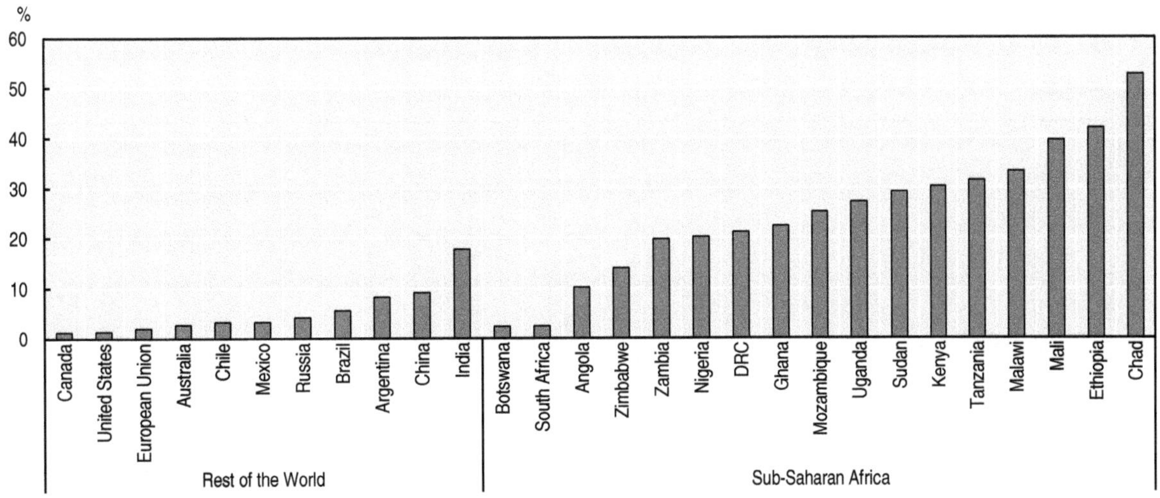

Note: DRC refers to Democratic Republic of the Congo.
Source: World Bank (2016).

StatLink http://dx.doi.org/10.1787/888933381341

These megatrends are not inevitable and remain subject to a degree of uncertainty as well as future policies. The focus on SSA as a whole risks perpetuating a view of Africa as a single entity, however this chapter aims to provide an agricultural outlook, with a wider consideration of links to growth and food security, that reflects the complexity within the region.

The agricultural environment in Sub-Saharan Africa

After decades of stagnation, much of Africa is now experiencing rapid economic transformation. In the post-structural adjustment period, the business environment has become more stable and albeit from a small base, the region has experienced rapid economic growth since the mid-1990s. These domestic factors in combination with the global "commodity boom" enabled commodity exporting countries in particular to achieve growth rates above or near 6%. However, the recent decrease in agricultural commodity prices, lower demand arising from China and currency depreciation have tempered growth rates in African economies. Foreign investment and external financial flows into Africa have quadrupled since 2000. These flows are expected to increase further in the coming years (AfDB, OECD & UNDP, 2014),[3] while internally generated funds in the form of tax revenues continue to rise across the continent.

Drivers of agricultural growth

Evident from its high share in GDP (Figure 2.1), the prospects of the agricultural sector heavily influence economic development in most countries in Sub-Saharan Africa. From 1990 to 2013, the total value of agricultural production, measured in constant US dollars, increased by 130% (Figure 2.2). The crop sector dominates total agricultural production value, accounting on average for almost 85% of total production value over the 24-year period. This share differs across the region, ranging from 53% in Southern Africa, to more than 90% in Western Africa.

Figure 2.2. **Gross agricultural production value in Sub-Saharan Africa**

Measured in constant 2004-06 US Dollars

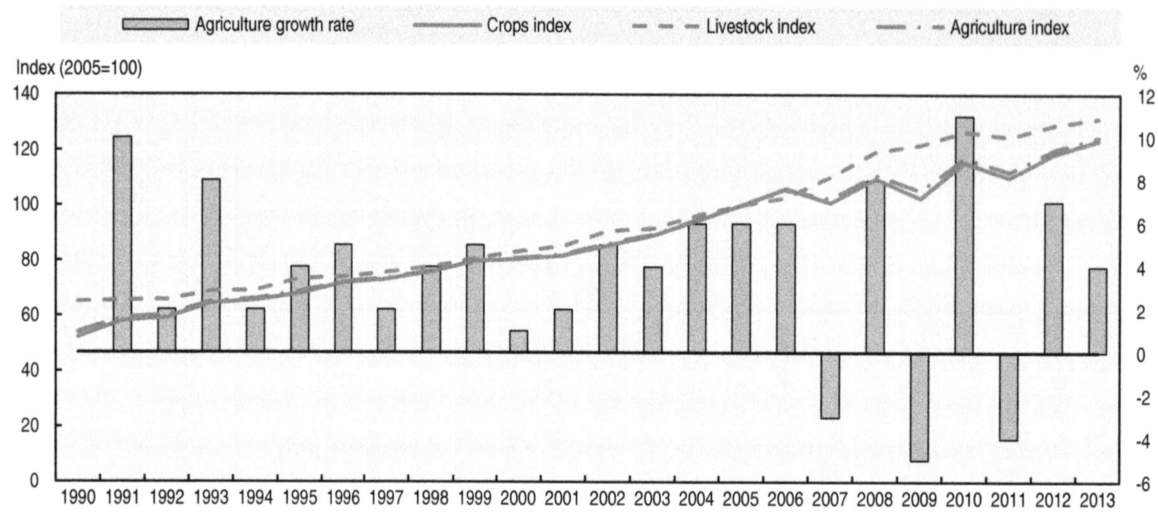

Source: FAOSTAT (2016), FAO, *http://faostat3.fao.org/*.

StatLink http://dx.doi.org/10.1787/888933381351

Regional differences in the relative contribution of the crop and livestock sub-sectors reflect agro-ecological and cultural diversity. Significant growth was evident across SSA over the past 24 years, but Western Africa continues to account for more than 60% of the total value of agricultural output in SSA, while Southern Africa contributes 22%. Western Africa has also been the greatest driver of volatility in total production value since 2007 (Figure 2.2), mainly due to volatile yam production in Nigeria. Despite the diversity in crop mix across the region, the crop sector's share of total production value is significantly higher in all sub-regions except Southern Africa, where the shares of livestock and crop production value are similar.

Within each of the four sub-regions, the five biggest crops contribute more than 45% of total crop production value, with maize being the single most important staple crop. Rice is an important staple in Eastern and Western Africa, and other important staples include potatoes (Eastern and Central Africa), sweet potatoes (Eastern Africa), cassava (Western and Eastern Africa) and plantains (Eastern and Central Africa). In Southern Africa, the strong share of fruits and vegetables in total value of production is due to South Africa's export oriented horticultural production.

The livestock production mix exhibits similar diversity, not only in its contribution to the total value of agricultural output, but also to the relative importance of the different livestock subsectors. Poultry contributes a substantial share of livestock production value across the region, ranging from 12% in Eastern Africa to 45% in Central Africa and Southern Africa (Figure 2.4). Interestingly in Central Africa, where livestock production value is smaller than any of the other three regions, game meat accounts for 35% of livestock value. Livestock production systems remain largely extensive, with pasture based ruminant production often the only system able to add value in semi-arid areas. Often the movement of livestock in line with seasonal changes and fodder availability remains the only way of securing feed for large herds (NEPAD, 2014). At the same time, vertically integrated, intensive poultry operations that link commercial feed grain producers to feed mills,

Figure 2.3. **Crop mix across the Sub-Saharan African region**

Based on average value of production 2011-13

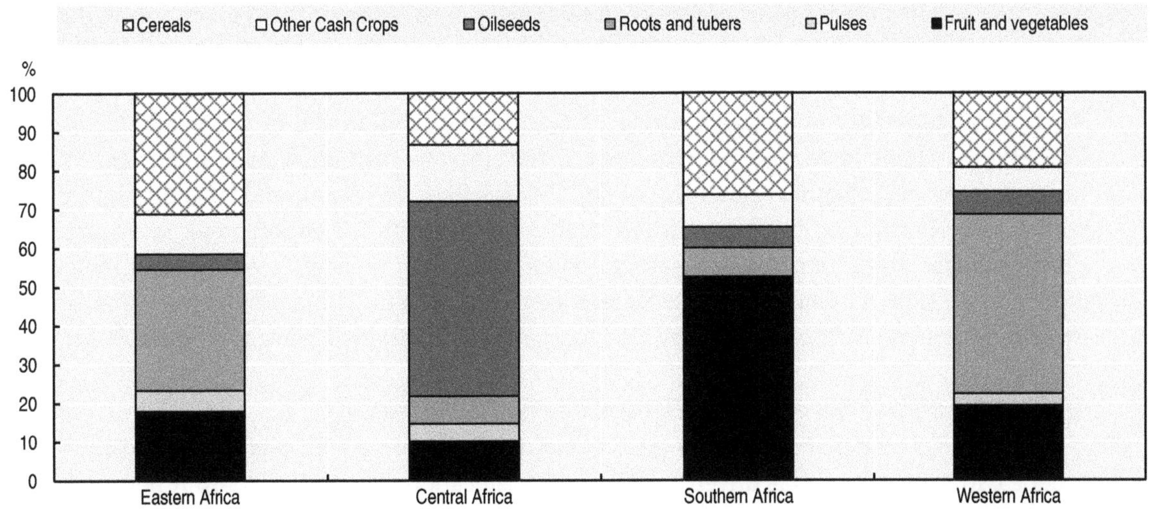

Source: FAOSTAT (2016). FAO, *http://faostat3.fao.org/*.

StatLink http://dx.doi.org/10.1787/888933381368

Figure 2.4. **Livestock mix across the Sub-Saharan African region**

Based on average value of production 2011-13

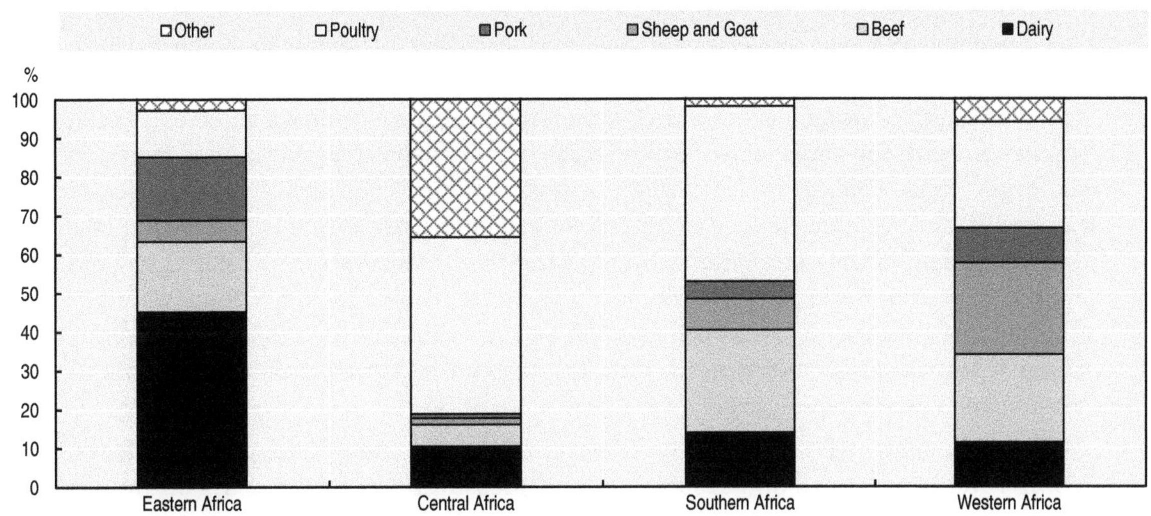

Source: FAOSTAT (2016), FAO, *http://faostat3.fao.org/*.

StatLink http://dx.doi.org/10.1787/888933381371

abattoirs and wholesalers have been evident for many years in some countries (such as South Africa) and have recently started expanding in others (such as Zambia and Tanzania). Such operations target urban consumption centres characterised by rapid growth in food demand.

Fisheries and aquaculture also make a multifaceted contribution to national economies in SSA. The region has vast fish resources, in marine and inland waters, and is characterised by diverse fishing communities. Small and industrial-scale fishing targets a diverse array of species for local and international consumption. Benefits generated by the sector include nutrition and food security, livelihoods, employment, and foreign revenue. Similarly to other agricultural sub-sectors, fisheries and aquaculture are considered to be underutilised, despite inherent potential. Total fishery production in the region represents only about 4% of world production and growth prospects are limited in the next decade. Fisheries and aquaculture in SSA faces enormous challenges and deficiencies which are limiting the capability of most governments to ensure its sustainability and profitability. Challenges include inadequate management of fish stocks; lack of knowledge and evidence to foster reforms; untapped potential of small-scale fisheries; weak and uncoordinated institutions; limited or ineffective institutional and legal frameworks; weak scientific research; inadequate human and financial resources; lack of reliable, relevant and timely information; lack of adequate infrastructure and services; and climate change, which is expected to change future fisheries production patterns, as species move to new habitats.

Agricultural growth has been underpinned by area expansion

The African model of agricultural growth differed significantly from that of Asia or South America. In Asia, growth was driven largely by intensification, whereas in South America, it was the result of significant improvement in labour productivity arising from mechanisation. By contrast, strong growth in SSA agricultural output has accrued predominantly from area expansion and intensification of cropping systems, as opposed to large-scale improvement in productivity (NEPAD, 2014; Brink and Eva, 2009). While the agricultural labour force has expanded, NEPAD (2014) notes that productivity per agricultural worker has improved by a factor of only 1.6 in Africa over the past 30 years, compared to 2.5 in Asia.

Given that SSA is generally regarded as land abundant, continued area expansion in the coming decade may not seem problematic. However rural SSA is highly heterogeneous and while much of its land is unutilised or underutilised, a considerable share of its rural population resides in smallholder farming areas that are densely populated and face land shortages (Jayne et al., 2014). In a wider assessment that considers a combination of biophysical and economic factors as criteria for viability, Chamberlin et al. (2014) indicate that potentially arable cropland is highly sensitive to assumptions related to land productivity and market access. Much of the underutilised land is concentrated in relatively few countries and between one half and two thirds of surplus land is currently under forest cover. Conversion of such forest land to agriculture would come at considerable environmental cost.

In land constrained countries, area driven growth may come at the expense of fallows. Rising rural populations and associated land pressures has resulted in continuous cropping in many African countries, with fallows largely disappearing in densely populated areas.[4] Continuous cultivation of existing plots would not necessarily pose problems for sustainable intensification if sufficient use of fertilisers, soil amendment practices and other land-augmenting investments are employed and coupled with continued education to maintain and improve soil quality. However, a large body of literature in SSA points to soil degradation arising from unsustainable cultivation practices in regions with a high population density, for example parts of Kenya and Malawi (Stoorvogel and Smaling, 1990; Drechsel et al., 2001; Tittonell and Giller, 2012). Continuous cultivation and lack of crop rotation deplete organic carbon levels, making soil less responsive to fertiliser application. This also makes it more difficult for smallholder farmers to benefit from yield gains offered by plant genetic improvement.

Agriculture's contribution to employment

The agricultural sector has a pivotal role in employment in SSA, employing more than half of the total workforce. While its importance to the rural population is well documented, recent surveys suggest that agriculture is also the primary source of livelihood for 10% to 25% of urban households. National census data indicates that the number of people employed primarily in agriculture has increased over time (Yeboah and Jayne, 2015).

As a consequence of rapid population growth, SSA has a young population, which will result in about 17 million people entering the labour force annually over the next decade (Losch, 2012; IMF, 2015). Given current employment growth rates, less than half will be absorbed into gainful non-agricultural employment and even under more favourable policy and growth scenarios, the share rises to two thirds at most (Fine et al., 2012). Consistent with employment trends by the Groningen Global Development Centre (2013), the World Bank reports that family farming will remain the single largest source of employment in the coming decade (Filmer and Fox, 2014). Thus youth participation in agriculture will largely depend on the viability of family farming.

Innovative ways of facilitating youth participation in agriculture have the potential to drive widespread poverty reduction among youths and adults alike. A coherent and integrated approach that addresses challenges related to education, land access and tenure, access to financial services, access to markets, access to green jobs and involvement in policy dialogue has the potential to make the agricultural sector more attractive to young people, providing the additional push that may be needed for them to enter the sector (FAO, CTA and IFAD, 2014).

The potential for urbanisation and income growth to stimulate job expansion in downstream segments of the food system also depends on where the primary agricultural products come from. Faster growth in domestic production arising from commercialisation could generate job growth in food assembly, wholesaling, processing and retailing. Box 2.1 provides further detail on the emergence of the West African food economy and the role of value addition in creating employment opportunities.

Box 2.1. **Emergence of a West African food economy**

West Africa's cities are now home to 133 million people, 25 times more than in 1950. Between 2000 and 2010 alone, the urban population grew by over 48 million people. Consequently, the size of the food economy grew spectacularly. Bolstered by urbanisation and income growth, household food consumption patterns are changing and the food economy is developing.

A growing number of households are turning to markets for their food supply. In urban areas, almost all food is bought on the market with an average of 93% of household food consumption supplied through various distribution channels. At the same time, increasingly diversified rural economies and the spread of urban products and lifestyles mean that the share of rural food supply from markets is also growing. Overall, markets provide at least two-thirds of household food consumption at the regional level (OECD, 2013).

Urbanisation and urban lifestyles are also accompanied by shifts in dietary patterns, which are spreading beyond the frontiers of towns and cities. More fruits, vegetables and processed foods are being consumed, while the share of cereals and pulses is declining. Urban consumers are clearly moving towards higher value food products. Fruits and vegetables, and meat and fish now account for half of the total food expenditure by urban households. The demand for convenience is an overarching trend across income groups and area. This is reflected in the strong demand for processed and prepared foods and in the expansion of street food. In urban areas, processed foods represent 41% of food budgets. Rural households, although less than their urban counterparts, still spend 36% of their budget on processed foods (Figure 2.5).

Box 2.1. **Emergence of a West African food economy** *(cont.)*

Figure 2.5. **West African food basket by food groups and area in 2010**

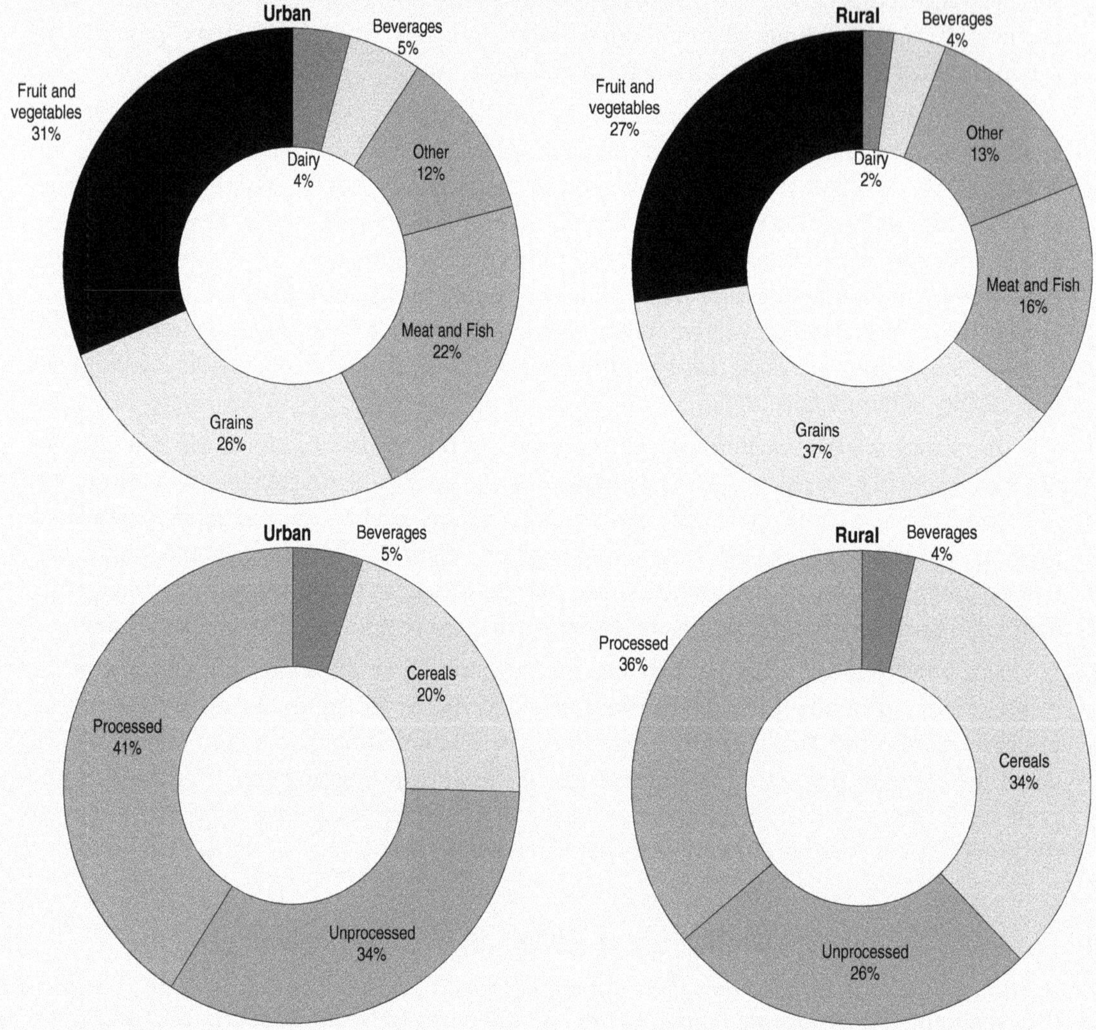

Source: Global Consumption Database, World Bank; SWAC/OECD.

StatLink http://dx.doi.org/10.1787/888933381386

The combined effects of rapid urbanisation, population growth and resulting transformations in food demand have had major impacts on the size of the West African food economy and its structure. Using recent expenditure and consumption surveys compiled by the World Bank, the size of the West African food economy is estimated at USD 178 billion in 2010.[1] This represents 36% of regional GDP, making it the largest sector of the West African economy. In many countries, the domestic food market is becoming more attractive for farmers than traditional export cash crops. The non-agricultural postharvest activities of the food economy, such as processing, logistics and retail, are developing quickly. These account for 40% of the sector's value added and will continue to expand (Allen and Heinrigs, 2016).

The emergence of local food industries and processing facilities creates increasing employment opportunities in processing, packaging, distribution and retail in urban and rural areas. In many rural areas the non-farm rural economy is growing rapidly and driving economic transformations.

Box 2.1. **Emergence of a West African food economy** *(cont.)*

To fully leverage the opportunities associated with urbanisation and the development of the food economy, a refocusing of agricultural policy towards an integrated food policy is needed. Supporting farmers in satisfying urban demand for fruits and vegetables, and meat and fish needs to be complemented by improving the business environment for agro-food businesses in processing, distribution and retail, and by improving the connections between urban areas and their rural hinterlands.

1. Calculated as the sum of all food consumption and adjusted by GDP figures.

Agricultural policies in Sub-Saharan Africa

Motivated by the need for a vibrant and sustainable agricultural sector, a number of policy initiatives have been integral to the sector's development over the past decade. CAADP was prioritised within the 2003 Maputo Declaration on Agriculture and Food Security through commitments to allocate at least 10% of national budgetary expenditure towards its implementation and aimed to achieve a 6% annual growth of the agricultural sector. Less than 20% of countries have achieved their commitment on agricultural spending. More recently, these commitments were reaffirmed in the Malabo[5] declaration on accelerated agricultural growth, which pledged to end hunger in Africa by 2025.

Despite the prioritisation of the agricultural sector, FAO's Monitoring and Analysing Food and Agricultural Policies (MAFAP) programme notes an overall decreasing trend in the share of public resources channelled to agriculture in the ten countries reviewed in 2013.[6] These countries have spent public funds on a broad range of consumer and producer support policies (Figure 2.6). However, some of these expenditures may have targeted primarily short-term objectives that may not have been fully aligned with long-term development goals for the sector (AfDB, FAO and ECOWAS, 2015). The strategic development of the agricultural sector would benefit from increased policy focus on

Figure 2.6. **Public expenditure on the agricultural sector in selected Sub-Saharan African countries**

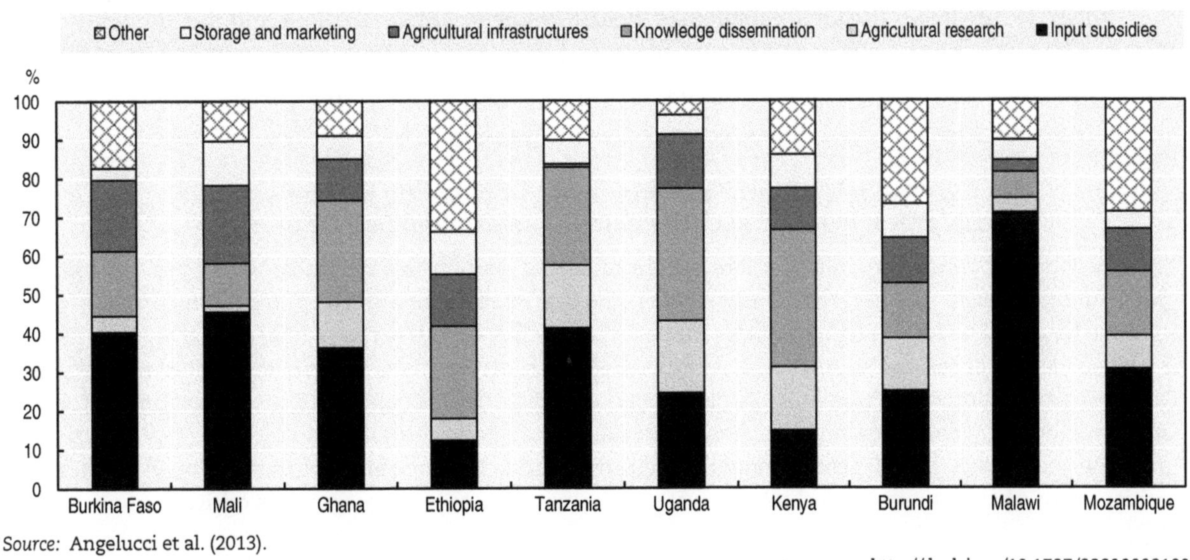

Source: Angelucci et al. (2013).

StatLink http://dx.doi.org/10.1787/888933381397

infrastructure, research and development. A lack of stability in the political and policy framework has been identified by numerous researchers as a constraining factor to the sector's development. Consistency of policy applications will remain a key factor shaping the success of the sector within the development agenda.

As a form of producer support targeting improved productivity, fertiliser subsidy programmes have been employed in a number of countries and while successful in accelerating yield growth in countries such as Zambia and Malawi, the ultimate effectiveness of such programmes in the long-run remains disputed, with the costs often found to outweigh the benefits (Jayne and Rashid, 2013). As an alternative, a holistic approach to support small-scale producers has been encouraged. This includes investment in agricultural R&D, extension programmes focused on improved soil quality and physical infrastructural development.

The establishment of strategic food reserve systems to support food security was a resolution within the Maputo declaration on agriculture and food security. Consequently most of the funds allocated to consumer related programmes in the region have been spent on maintaining public food stocks of important staple grains. The implementation costs, as well as the price distorting impacts of such policies are important considerations. They have been minimised where their application is based on transparent target prices that are in line with import and export parity levels. Additionally, governments used temporary trade policies such as export bans or import tariff reductions to support consumers.

Production growth in SSA has failed to keep pace with demand deriving from population and income growth, resulting in rising imports for food commodities such as wheat, rice and poultry. In many instances, import tariffs have been employed to support domestic producers, particularly relative to other producers outside the region. Box 2.2 illustrates some of the possible benefits from increased intra-regional trade in Eastern and Southern Africa. Multiple regional trade agreements in Africa, such as the Common Market for Eastern and Southern Africa (COMESA), the East African Community (EAC), the Economic Community of West African States (ECOWAS) and the Southern African Development Community (SADC), have been successful in reducing tariff rates, but have also been accompanied by a commensurate rise in non-tariff measures. Implementation of the tripartite free trade area, established in 2015 between SADC, COMESA and the EAC will result in the largest economic bloc on the continent, covering more than 50% of Africa's population and GDP, it has the potential to impact significantly on trade in the region.

Box 2.2. **The role of intra-regional trade in reducing market volatility and improving food security across Eastern and Southern Africa**

High domestic food price volatility has been a recurring issue in many African countries. It poses a particular risk to the food security of poor households that spend a greater share of their income on food and for households depending on agriculture for their livelihood. The perceived need by governments to stabilise volatile prices and supply has long motivated their interventions in agricultural markets, despite the international drive towards liberalisation (Jayne and Tschirley 2009, Minot 2014). In reacting to these market shocks, governments are often faced with the need to balance short-term food security objectives with the longer term goal of raising productivity growth. In SSA, the most food insecure region in the world, achieving this balance remains a challenge, yet consistent policy application remains critical.

Box 2.2. **The role of intra-regional trade in reducing market volatility and improving food security across Eastern and Southern Africa** *(cont.)*

The role of intra-regional trade in reducing volatility and improving food security was recognised by the African Union (AU) in its Malabo Declaration on accelerated agricultural growth, which committed to boosting intra-African trade in agricultural commodities and services. Reductions in barriers to regional trade offer an inexpensive means of reducing domestic prices and hold enormous potential to improve food security in the region. This positive contribution is already evident in regions where neighbouring countries are pooling production to stabilise markets through cross border trade (Mozambique-Malawi, Malawi-Zambia, Uganda-Kenya) (Chapoto and Sitko, 2014).

The emergence of Zambia as a consistent surplus producer, particularly for maize, has resulted in changing regional trade patterns in Eastern and Southern Africa. A favourable transport differential and the absence of genetically modified (GM) technology have made it the preferred source for Zimbabwean imports. At the same time, discretionary trade restrictions from the Zambian government have influenced the consistency of supply into Zimbabwe, impacting on price volatility. Over the course of the Outlook projection, the extent to which Zambia continues to grow as an exporter into Zimbabwe and other deficit countries in the region will depend on the consistency with which its trade policies are applied. South Africa and Uganda have maintained open trade policies and are projected to continue supplying exports consistently into the region. Assuming that borders remain open, Zambia is also projected to expand exports, becoming the second largest maize exporter in SSA in the coming decade (Figure 2.7).

While intraregional trade in maize is higher, it still accounts for less than 10% of total trade in food staples. Informal trading charges, burdensome border regimes and limited transportation infrastructure have all been identified as impediments. Priority should be given to actions and investments that reduce these impediments and put in place more predictable rules-based systems (Morrison and Sarris, 2016).

Figure 2.7. **Net trade of maize in Eastern and Southern Africa**

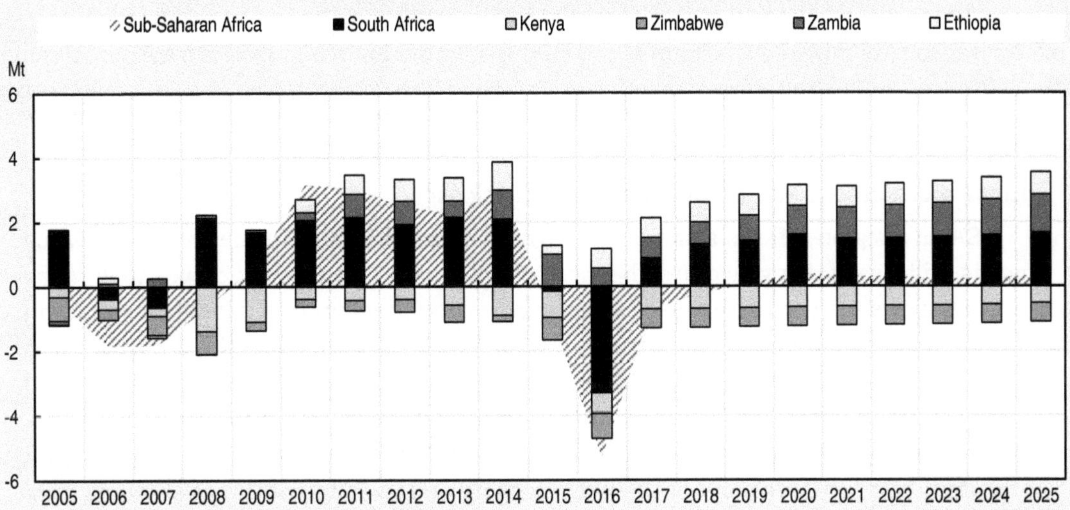

Note: The shaded area in the figure illustrates the aggregated net trade position for maize in SSA, and the stacked bar charts illustrate the net trade position of the largest importers and exporters in ESA.

Source: OECD/FAO (2016), "OECD-FAO Agricultural Outlook", OECD Agriculture statistics (database), *http://dx.doi.org/10.1787/agr-data-en*.

StatLink http://dx.doi.org/10.1787/888933381402

Key factors underlying the agricultural outlook for the region

The prospects for production, domestic demand and trade of agricultural commodities in the region are influenced by a host of agro-ecological, economic, demographic and political factors. The level of income available in the domestic food market is driven by general economic development in the countries. The prospects of several countries that are highly dependent on commodity exports have declined and per capita GDP growth is projected to stagnate over the coming decade, resulting in a significant slowdown in the Central, Southern and Western African regions (Figure 2.8). Accelerated output growth in the Eastern African region offsets some of the decline however and GDP per capita for the entire SSA region expands only marginally more slowly in the coming decade (2.3% p.a.) relative to the past (2.4% p.a.).

Figure 2.8. **GDP growth per person in Sub-Saharan Africa**

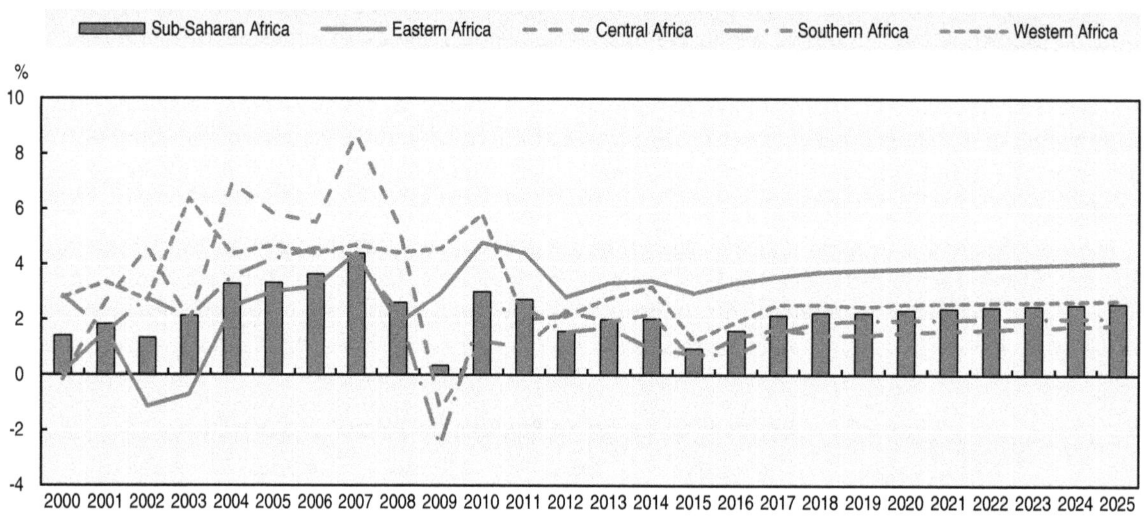

Source: OECD/FAO (2016), "OECD-FAO Agricultural Outlook", *OECD Agriculture statistics* (database), *http://dx.doi.org/10.1787/agr-data-en.*

StatLink http://dx.doi.org/10.1787/888933381419

Income distribution: Rise of an African middle class?

SSA as a region reflects some of the highest inequality rates in the world and while there is evidence of rising incomes, the increasingly skewed distribution of wealth in many countries slows the rise of middle class consumers and thereby constrains the increase in domestic food expenditures. The African Development Bank (AfDB) defined middle class as per capita daily consumption of USD 2-20 in 2005 purchasing power parity terms and indicates that its share of the population has expanded from 27% in 1990 to 34% in 2010. However, roughly 60% of the middle class in 2010 were in the USD 2-4 per capita consumption group–barely out of the poor category and in constant threat of falling back into it (AfDB, 2011).

Demographic structure

Population growth within SSA has significantly outpaced any other region in the world. Since 1990, the SSA population expanded by 96%, more than double the world average of 38% (45% in Oceania, 37% in Asia, 27% in North America and less than 3% in Europe). Over the next decade, a further expansion of 28% is projected, compared to a global average of only 11%. The rapid expansion has resulted in a unique demographic structure and more than 60% of the population is below the age of 25, as opposed to 41% in Asia and only 27% in Europe. The economic potential of demographic dividend, the time period during which the share of the working-age population is larger than the non-working-age share, should be acknowledged.

The share of the population residing in urban areas has increased to 38% in 2015, from 27% in 1990, a rate similar to South America and Southern Asia. By 2025, it is projected to increase to 42%, impacting on income levels and dietary patterns. Despite urbanisation, the rural population has continued to increase in absolute numbers and surveys indicate that even in urban areas agriculture still represents the primary livelihood for up to 25% of the population. Enhanced productivity in agriculture therefore appears to have the greatest potential to directly improve rural livelihoods, while stimulating effective demand and job opportunities in the nonfarm sector through multiplier effects generated from productivity gains.

Emerging medium-scale producers

The surge in global food prices post 2007, combined with agricultural subsidies and land policies in many countries accelerated the demand for agricultural land in SSA. Land acquisitions by foreign and African investors have increased dramatically leading to the rapid rise of medium- and large-scale "emergent" commercial farms (Jayne et al., 2015). Development Health Survey data indicates that urban households now control 10-30% of total agricultural land. Evidence also suggests that existing land policies are leading to increased inequality of landholdings and in some cases may be making it more difficult for area expansion in densely populated smallholder farming areas.

The relative productivity of these medium-scale farmers is less clear. Farm businesses may benefit from increased commercialisation and economies of scale, but speculative land acquisitions by richer, politically influential urban households may not impact on productivity. Increased commercialisation is projected to raise productivity growth, yet a substantial yield gap remains. Accelerated changes to farm structure, accompanied by mechanisation and improved farming practices have the potential to induce a much higher rate of productivity growth.

Medium-term outlook

Focusing on the most food insecure region in the world, key questions for the medium-term outlook relate to growth in caloric intake. Significant expansion in the demand for food products is inflated by population growth, and improvements in per capita caloric intake remain modest. Regional differences are evident both in levels and composition, but the highest growth is achieved in Eastern Africa, where per capita caloric intake expands by almost 7.5% (162 kcal/day/person) by 2025. By contrast, the Central African region, heavily influenced by its two key countries, Angola and the Democratic Republic of the Congo (hereafter "DRC"), increase caloric intake per capita by only 1.5% (32 kcal/day/person) over the ten-year period. This results from rapid population growth in excess of 3.5% in DRC and a reduced income growth outlook in Angola, which relies heavily on crude oil exports. Southern and Western Africa houses the largest economies in the region (South Africa and Nigeria), and therefore unsurprisingly, total caloric intake is almost 40% higher relative to Central and Eastern Africa. Significant diversity is still evident within these regions however and caloric intake in several countries remains low.

As a group, caloric intake in Southern and Western Africa will be 6% (178 kcal/day/person) and 5% (136 kcal/day/person) higher respectively by 2025, with the bulk of the increase coming from vegetable oil and sugar consumption (Figure 2.9).

Figure 2.9. **Caloric intake by commodity group in Sub-Saharan Africa**

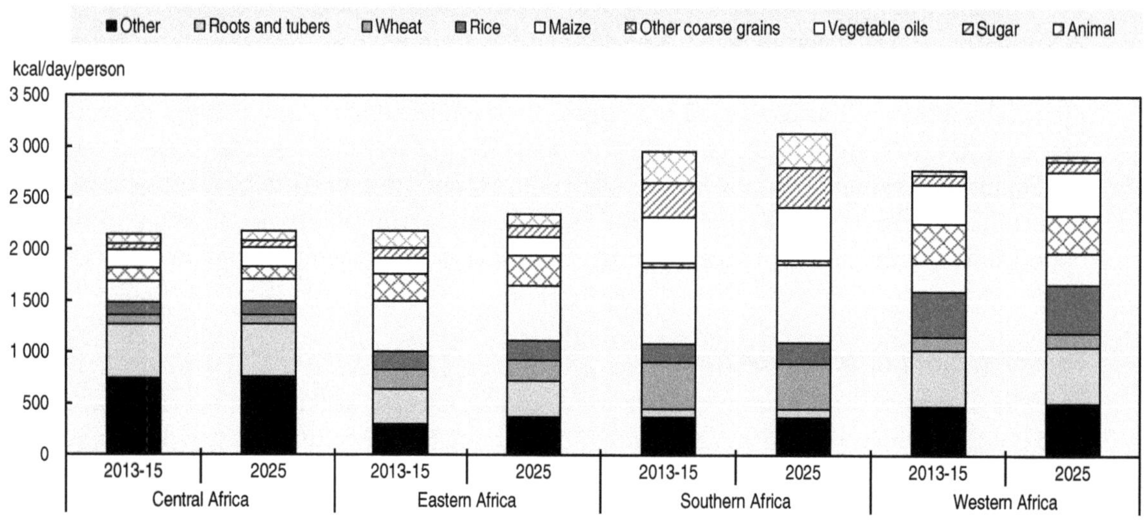

Source: OECD/FAO (2016), "OECD-FAO Agricultural Outlook", *OECD Agriculture statistics* (database), *http://dx.doi.org/10.1787/agr-data-en.*
StatLink *http://dx.doi.org/10.1787/888933381429*

Per Capita protein intake also differs across the region, being almost 65% higher in Southern Africa relative to Central Africa (Figure 2.10). This reflects significantly higher meat consumption in Southern Africa. In line with meat and dairy, per capita protein consumption grows fastest in Eastern Africa, expanding by 6% (4 g/day/person) over the ten year period. In Central Africa, which remains beset by civil conflict, per capita growth is modest at just over 3% (2 g/day/person) for the ten-year period.

Figure 2.10. **Per capita protein consumption by commodity group in Sub-Saharan Africa**

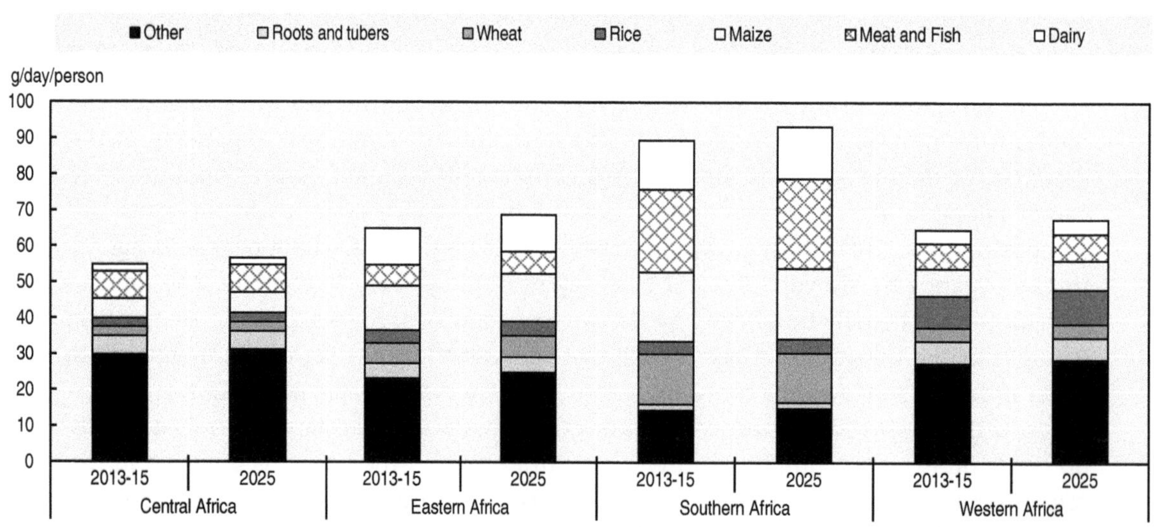

Source: OECD/FAO (2016), "OECD-FAO Agricultural Outlook", *OECD Agriculture statistics* (database), *http://dx.doi.org/10.1787/agr-data-en.*
StatLink *http://dx.doi.org/10.1787/888933381434*

Combined with a rapidly expanding population, modest increases in per capita consumption imply significant demand growth and rising import demand for many commodities. Figure 2.11, which shows indices of the value of agricultural production based on constant 2004-06 US dollar prices and production projections, suggests that production also responds. This response is supported by high prices in the SSA region resulting from both weaker exchange rates and price premiums resulting from domestic market isolation; yet significant differences remain between different sub-regions and commodities. Based on stable yield assumptions, the region remains self-sufficient in key staples such as maize, and roots and tubers, with imports only required in times of drought. By contrast, a significant share of consumption growth for wheat, rice and poultry will be met through imports. At the same time, export industries such as sugar, cotton, fruits and beverages continue to grow, contributing to foreign currency reserves.

Figure 2.11. **Agricultural production index for covered commodities in Sub-Saharan Africa**

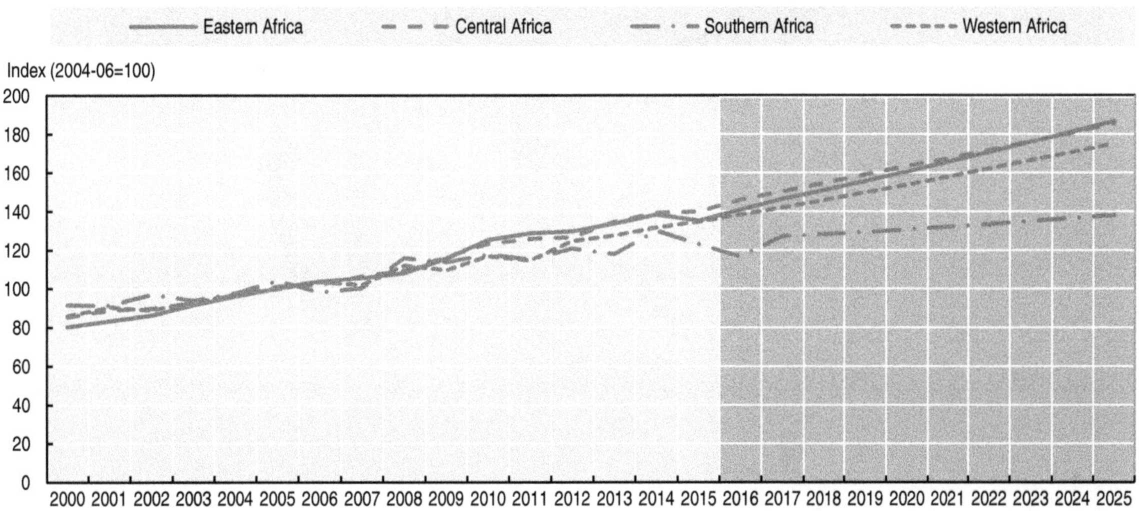

Source: OECD/FAO (2016), "OECD-FAO Agricultural Outlook", *OECD Agriculture statistics* (database), *http://dx.doi.org/10.1787/agr-data-en*.

StatLink *http://dx.doi.org/10.1787/888933381440*

Crops

Following rapid expansion over the past decade, adverse weather conditions across Eastern and Southern Africa reduced cereal output by more than 10% in 2015 (Box 2.3). Production prospects for the 2016 summer crop remain circumspect across Southern Africa in particular and prices have risen sharply in response. Maize prices in South Africa, Zambia and Malawi reached record levels early in 2016 and governments responded by releasing maize to consumers at subsidised prices and limiting export permit allocations.

Box 2.3. **Impact of the 2015-16 drought on agricultural production in Eastern and Southern African**

The SSA region has been particularly prone to recurrent drought conditions in the past. Between 1990 and 2013, almost 43% of the drought events recorded in the EM-DAT[1] database occurred in SSA. The impacts of precipitation shortages on agricultural output have been particularly severe, due to the predominance of rain-fed cropping and pasture based livestock systems. Climate projections suggest that rainfall variability is likely to increase in the SSA region over the coming decade, impacting on food security.

Box 2.3. **Impact of the 2015-16 drought on agricultural production in Eastern and Southern African** *(cont.)*

In 2015, the occurrence of a strong *El Niño* episode has been accompanied by exceptionally dry conditions across Eastern and Southern Africa (ESA). Ethiopia recorded the lowest annual rainfall in 30 years in 2015, while the same year represented the lowest annual rainfall since 1904 in South Africa. The monthly distribution of rainfall is an equally important consideration for agricultural production. The limited and uneven distribution of rainfall through the optimal planting period for summer crops (October to December) across the Southern African region had a particularly adverse impact on early crop development, raising food security concerns. The agricultural stress index in Figure 2.12 illustrates the extent of developing crops suffering from water stress as of late December 2015.

Figure 2.12. **Agricultural stress index – December 2015**

□0-40 ■40-70 ■70+ ■Off season

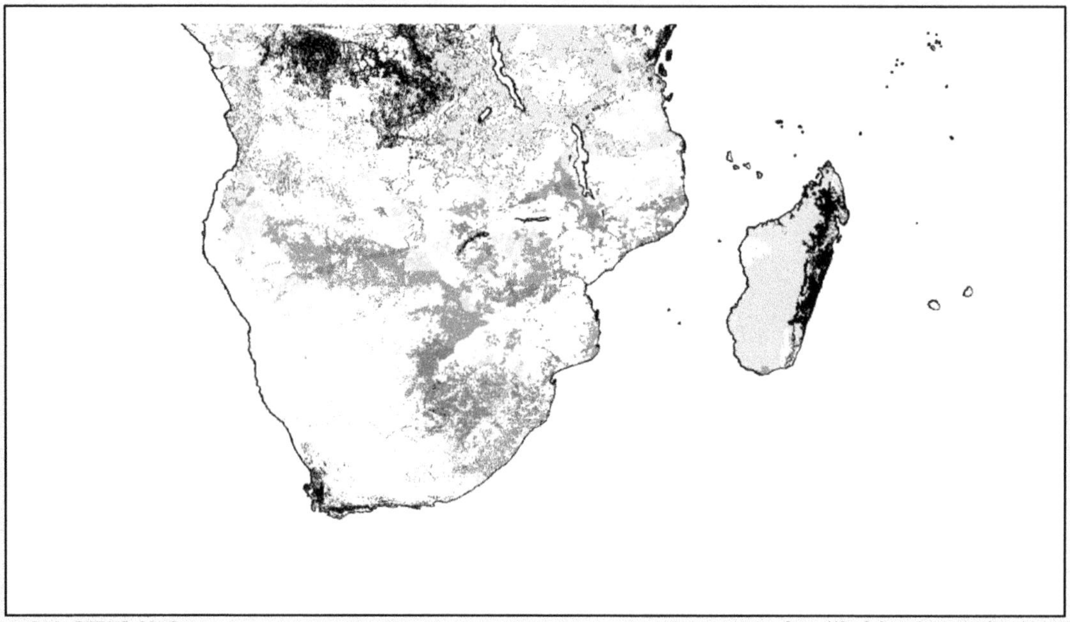

Source: FAO-GIEWS, 2016. StatLink http://dx.doi.org/10.1787/888933381459

Maize is the principal staple in most countries within ESA, hence its availability and affordability have been central to food security in the region. It represents the largest summer crop across most of ESA and provides a livelihood to multitudes of small scale producers. In South Africa and Zambia, the largest surplus producers in the region, initial production estimates reflect a decline of 27% and 21% in the maize crop in early 2016, from an already below average crop in early 2015. Consequently, import volumes across ESA will rise significantly in 2016 and, contrary to historic trends, the bulk will originate from outside the region.

Import bills are rising further as many currencies in the region have devaluated considerably against the US dollar. Furthermore, stringent regulations related to genetically modified (GM) technologies[2], as well as the preference for white maize limits potential procurement options,[3] incurring a substantial price premium over yellow maize, which is more commonly traded in the global market. Maize prices are projected to remain high until March 2017, which represents the first opportunity for early deliveries from within the region to alleviate pressure on low stock levels.

1. International disaster database – Centre for Research on the Epidemiology of Disasters: *www.emdat.be/database*.
2. Regulations related to GM technology vary across the region. The bulk of countries do not accept GM maize, while some countries are able to import it only under specified conditions.
3. There are only a few white maize surplus producers in the world of which Mexico and the US seem to be the only viable sources of white maize for the export market in the current marketing season.

The Zambian Food Reserve Agency stopped exporting its existing stock. Significant imports will be required to ensure short-term food security and in many coastal regions, wheat is being imported at more competitive prices than white maize.

Expansion of total crop area in SSA is projected to slow relative to the past decade due to a lower price environment, combined with the rising cost of bringing additional arable land into production (Figure 2.13). The bulk of additional land is allocated to staple crops such as coarse grains, rice, and roots and tubers, though the rates of expansion vary widely across regions and countries. In Southern Africa for instance, soybeans account for the greatest share of additional area, whereas in Eastern Africa it is coarse grains. In Central and Western Africa, the greatest increase is attributed to rice, and roots and tubers.

Figure 2.13. **Crop area in Sub-Saharan Africa**

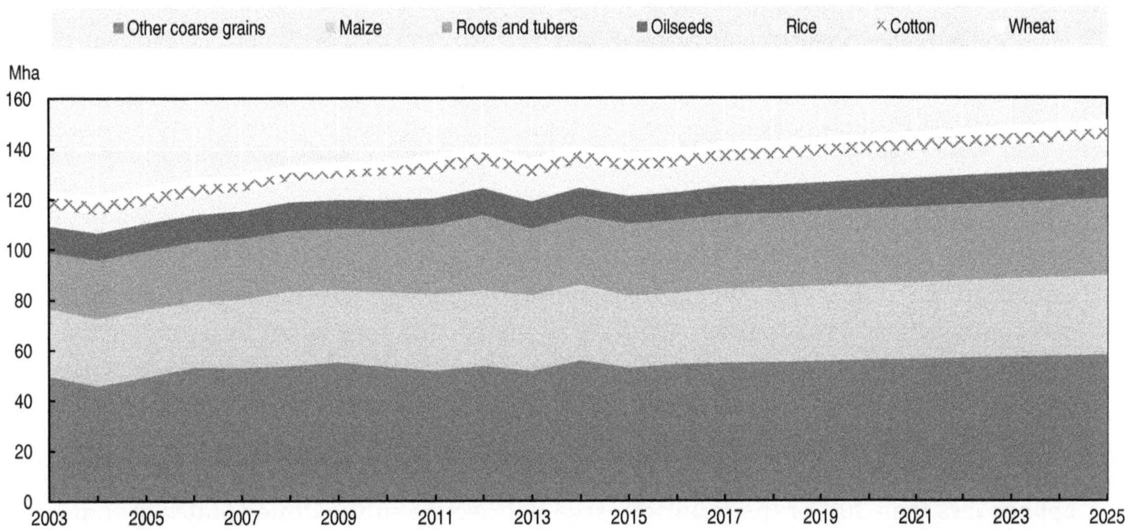

Source: OECD/FAO (2016), "OECD-FAO Agricultural Outlook", *OECD Agriculture statistics* (database), *http://dx.doi.org/10.1787/agr-data-en.*

StatLink *http://dx.doi.org/10.1787/888933381469*

Cereals

Cereals remain the primary source of energy for more than 962 million people across SSA and are therefore critical to food security. The composition of cereal consumption differs from the global norm however in that maize is an important staple for human consumption. Total cereal demand growth slows to 2.7% p.a. over the outlook period from 3.8% p.a. over the past decade; yet by 2025, total consumption will have increased by more than 52 Mt, just over 6kg per capita. By 2025, total cereal consumption will exceed 134 kg per capita, which remains less than 40% of the global average.

Food use continues to drive demand growth for all cereal products (Figure 2.14). Feed use is also an important driver of additional maize demand. By 2025, food consumption accounts for almost 70% of total maize demand, with an even higher share for other coarse grains, wheat and rice. Maize continues to dominate the cereal market, accounting for almost 40% of total cereal consumption by 2025, followed by other coarse grains (27%), rice (18%) and wheat (15%). In line with unique historic preferences, consumption growth differs by region, however, and while maize accounts for the largest share of additional cereal demand in Southern, Eastern and Central Africa, demand growth for rice exceeds any other cereal in West Africa.

Figure 2.14. **Cereal demand composition in Sub-Saharan Africa**

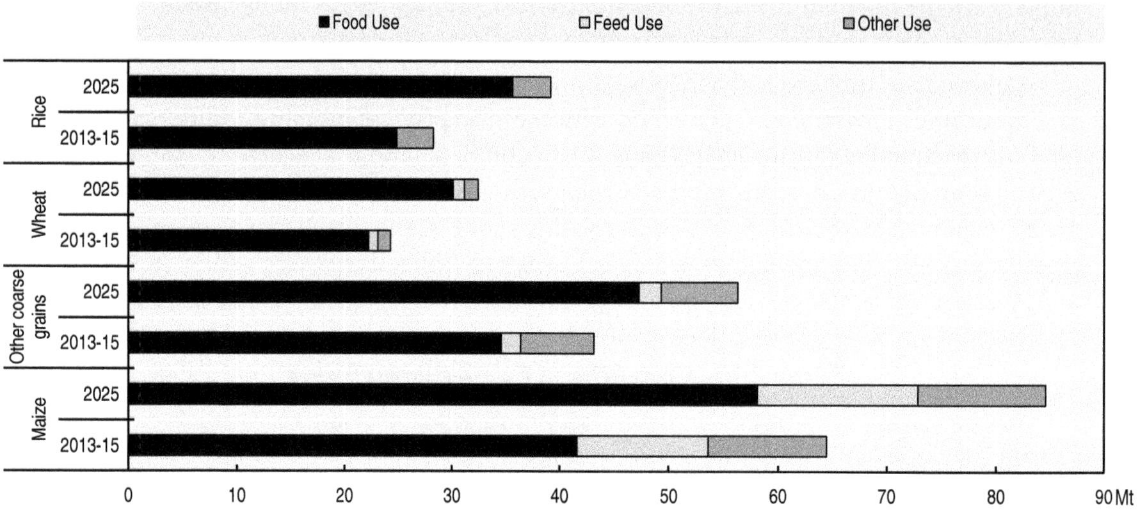

Source: OECD/FAO (2016), "OECD-FAO Agricultural Outlook", *OECD Agriculture statistics* (database), *http://dx.doi.org/10.1787/agr-data-en.*

StatLink *http://dx.doi.org/10.1787/888933381472*

Cereal production is projected to expand by 3.2% p.a. by 2025, rising by more than 41 Mt relative to the 2013-15 base period. Coarse grains account for the bulk of the expansion, with maize and other coarse grains contributing more than 80% of additional cereal production between them. Contrary to the past decade, when rising production resulted from area expansion, the role of productivity growth is expected to increase. Regional differences exist, but total area cultivated to cereals expands by about 1% p.a. over the coming decade, whereas yields improve by an annual average of just under 2% (Figure 2.15). With a few exceptions, fertiliser use remains well below global norms; SSA applies less than 20% of the fertiliser usage per hectare in the United States or India, suggesting that it could increase usage to support further yield growth. Yield growth could also be supported by the development of irrigation and the use of advanced seeds. Meaningful increases in fertiliser usage will however depend on resource availability, as well as the extent to which infrastructural development improves distribution to increase accessibility and reduce costs.

Maize production growth remains centred in a few countries; Eastern Africa will account for more than 50% of the additional 19 Mt of maize produced in SSA by 2025, and six countries together contribute more than 60% of the additional production (Figure 2.16). The relative contribution of land to production growth differs widely across the region, though robust yield growth of more than 1.8% p.a. is projected in all six countries. In South Africa, yield growth is sufficient to support higher production despite declining area, whereas in Ethiopia, significant expansion is evident in both area and yields. Albeit from a smaller base, production growth in Zambia and Uganda exceeds 40% over the outlook period, underpinned by continued area expansion coupled with yield improvements. Supported by input support programmes that enhance the accessibility of modern inputs to multitudes of small-scale producers, both countries produce a rising exportable surplus over the ten-year period.

Growth in production of other coarse grains is concentrated in Eastern and Western Africa, where sorghum and millet are popular within the cereal consumption basket. Between them, Eastern and Western Africa constitute more than 90% of production

Figure 2.15. **Change in area and yield for cereals in Sub-Saharan Africa**

Annual growth between 2016 and 2025

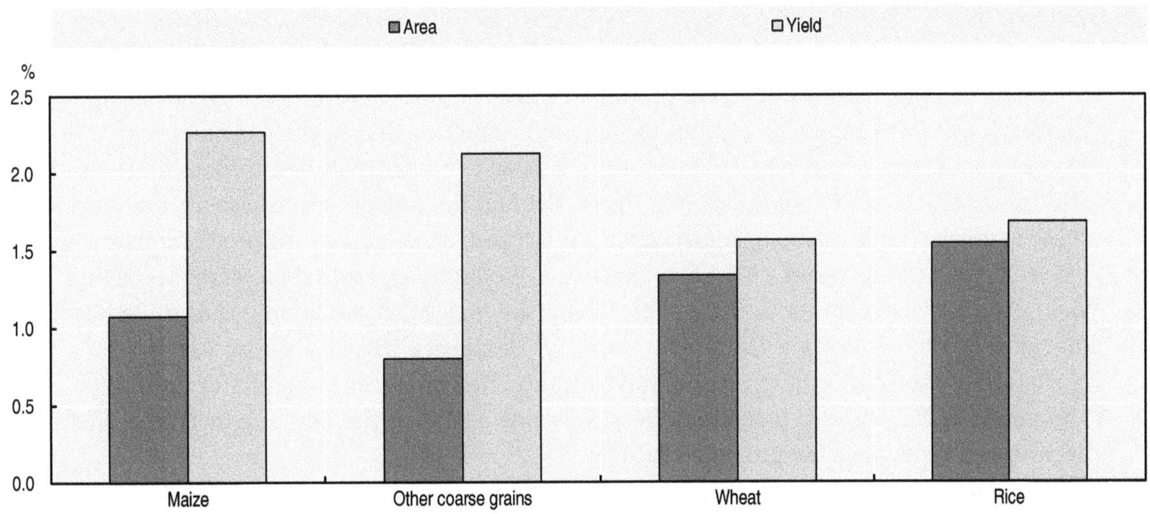

Source: OECD/FAO (2016), "OECD-FAO Agricultural Outlook", *OECD Agriculture statistics* (database), *http://dx.doi.org/10.1787/agr-data-en.*

StatLink http://dx.doi.org/10.1787/888933381486

Figure 2.16. **Maize production in selected Sub-Saharan African countries**

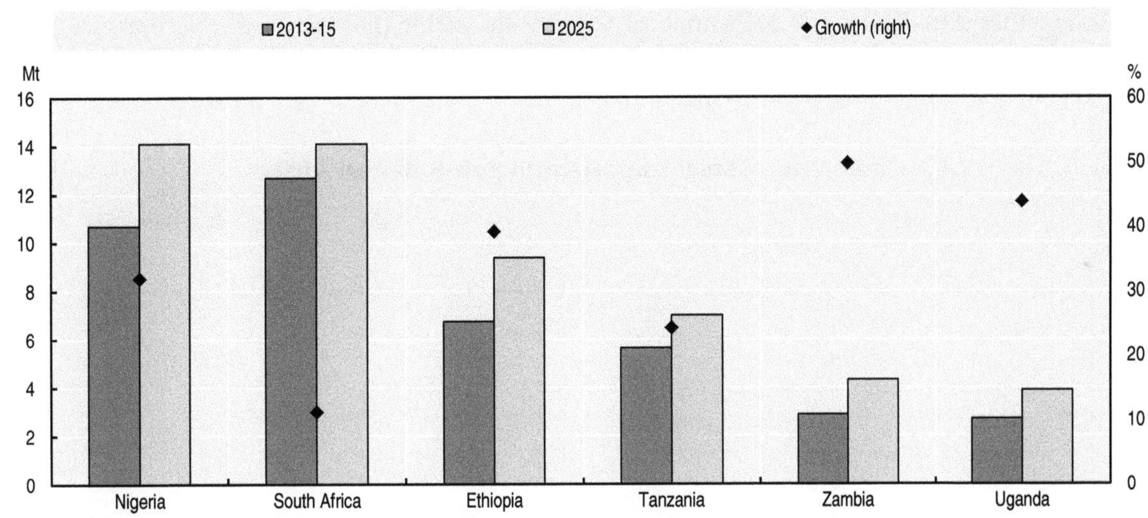

Source: OECD/FAO (2016), "OECD-FAO Agricultural Outlook", *OECD Agriculture statistics* (database), *http://dx.doi.org/10.1787/agr-data-en.*

StatLink http://dx.doi.org/10.1787/888933381497

growth. Ethiopia alone accounts for almost 40% of additional production to 2025, followed by Nigeria (14%) and Sudan (10%). Significant yield improvements are projected in all three countries, but the sharp production increase in Ethiopia and Sudan is further supported by area expansion of 18% and 22% respectively by 2025.

Less than 3% of global rice is produced in SSA, but at 6% p.a., its production has expanded faster than any other cereal over the past decade. Favourable storage characteristics, ease of preparation and versatility in consumption make rice a popular choice among consumers and by 2025, SSA will produce more than 20 Mt. The five biggest producers (Nigeria, Madagascar, Mali, Tanzania and Guinea) account for almost 65% of production growth. Area expands in all five countries, yet focused efforts by the African

Rice Centre and the International Rice Research Institute, combined with the adoption of improved varieties, such as locally developed New Rice for Africa cultivars also support yield growth.

From a production perspective, wheat is the smallest of the cereal markets in SSA, contributing only 5% of total cereal production in 2015. Few countries are endowed with the biophysical attributes for competitive wheat production and in the base period (2013-15), four countries accounted for more than 90% of the region's wheat production with Ethiopia accounting for more than 70%. The bulk of production growth is also attributed to these countries, with Ethiopia, South Africa, Sudan and Kenya expanding production by 4% p.a., 1.3% p.a., 1.9% p.a. and 2.4% p.a. respectively. Production growth results from area and yield expansion, except for South Africa, where the area of wheat is projected to decline marginally. This follows a long-term trend of declining wheat area in South Africa (particularly the Free State province) following the deregulation of agricultural markets. The sharp decline in area was offset by yield gains and over the Outlook, both the area decline and yield gains are projected to slow.

Despite impressive production growth, demand is such that cereal imports continue rising and by 2025 exceed 49 Mt – growing 2.2% p.a. In line with past trends, wheat and rice contribute the bulk of additional cereal imports, with coarse grains representing only 6% of the total growth. Wheat and rice imports expand by an annual average of 2.7% and 3.3% respectively. Imports are concentrated in a few countries; Sudan and Nigeria account for more than 20% of wheat and rice imports respectively. Within these products, all countries in the region remain in deficit and with few exceptions for rice, net imports rise across the region over the coming decade (Figure 2.17).

Figure 2.17. **Cereal imports into Sub-Saharan Africa**

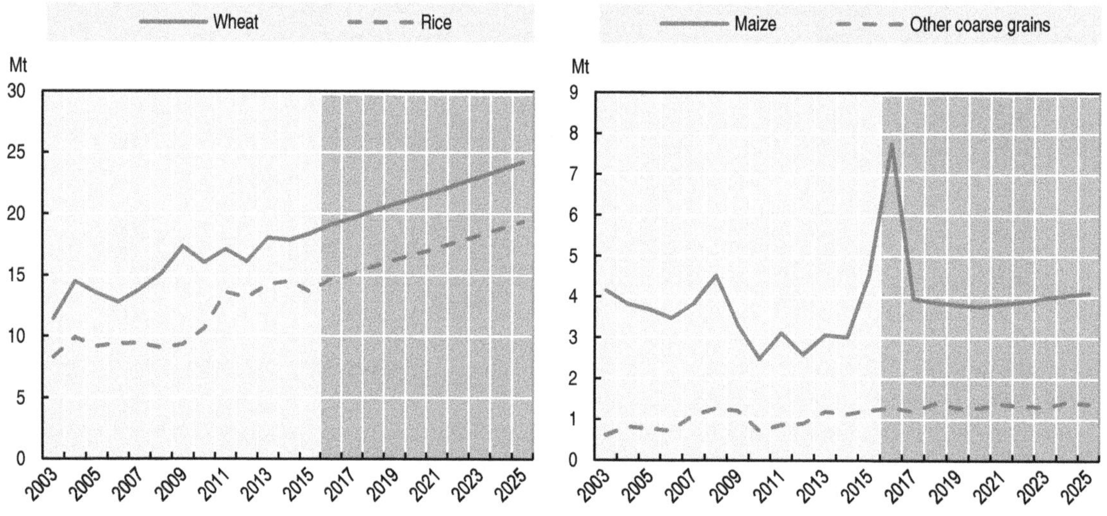

Source: OECD/FAO (2016), "OECD-FAO Agricultural Outlook", *OECD Agriculture statistics* (database), *http://dx.doi.org/10.1787/agr-data-en.*

StatLink http://dx.doi.org/10.1787/888933381509

The SSA region is a surplus coarse grain producer, and while the size of the surplus is projected to fall by 2025, the trade balance for both maize and other coarse grains remains positive. Contrary to wheat and rice imports, maize trade is predominantly intra-regional. Traditional surplus producers such as South Africa, Zambia and Ethiopia continue to account for the greatest share of export growth, while Kenya and Zimbabwe remain the largest deficit markets (Box 2.2). In South Africa, trade into the SSA region is projected to decline, as continued demand growth from the animal feed market supports a shift in production from white to yellow maize, resulting in surpluses of yellow maize entering the global market as opposed to the rest of the SSA region.

Roots and tubers

As an affordable and nutritionally rich staple, roots and tubers are an important constituent of SSA diets, particularly in Central and Western Africa, where per capita consumption exceeds any cereal product. Preferences differ regionally, but food consumption remains the primary component of total demand and products are sometimes blended with other imported staples such as wheat flour to curtail high food costs. Having surpassed 65 kg per capita, almost double the global average, per capita consumption in Central and Western Africa stagnates over the projection period but growth on a per capita basis occurs in Eastern Africa (Figure 2.18). Accounting for population growth however still results in robust expansion of total demand in SSA as a whole, approaching 100 Mt (55 kg per capita) by 2025. Of the additional 18 Mt consumed by 2025, almost 9 Mt is attributed to Western Africa, reflecting an average annual growth rate of 2.6%, compared to 2.4% and 2.8% in Central and Eastern Africa respectively.

Figure 2.18. **Roots and tubers consumption in Sub-Saharan Africa**

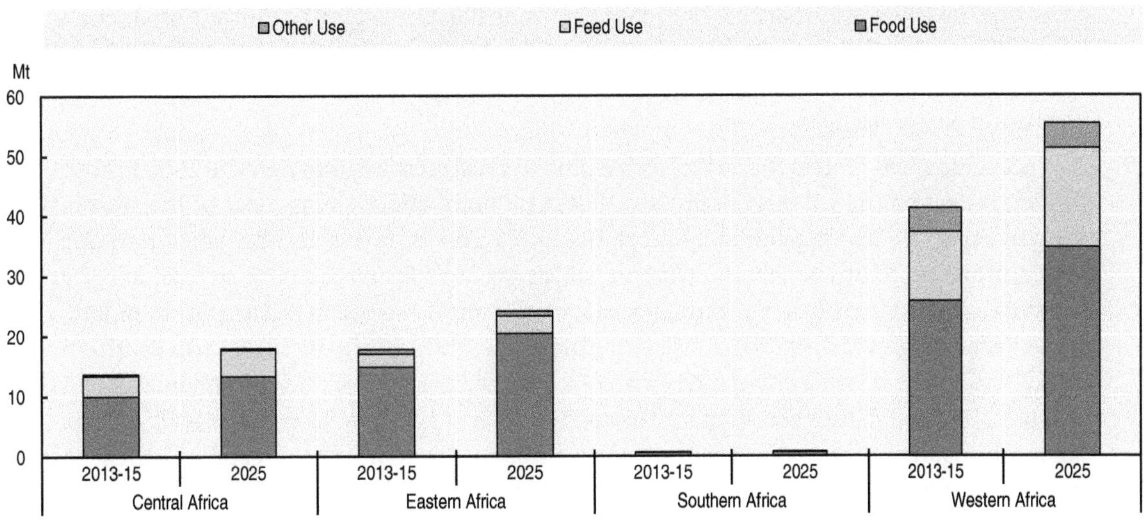

Source: OECD/FAO (2016), "OECD-FAO Agricultural Outlook", *OECD Agriculture statistics* (database), *http://dx.doi.org/10.1787/agr-data-en.*
StatLink *http://dx.doi.org/10.1787/888933381519*

Adaptation to marginal environments and flexibility in mixed farming systems allows roots and tubers to make a meaningful contribution to household food security and income levels, particularly for the rural poor. Given their perishable nature, trade represents a very small share of the market and production concentrated in Western Africa reflects consumption preferences. Nigeria alone accounts for more than 37% of production growth over the projection period, owing to yield gains that accelerate from the past decade. Different approaches have been developed to improve production technology for small scale producers and yield growth is supported by international partnerships designed to improve the adoption of new varieties.

Oilseeds and oilseed products

In line with global trends, oilseed production in SSA has expanded rapidly over the past decade, but was concentrated in a few countries. Soybean production soared by just over 1 Mt, yet almost 90% of the additional area was in South Africa, implying a total expansion of just over 0.1 Mt across the rest of SSA. Similarly almost 75% of other oilseed production growth is attributed to Nigeria, South Africa and Senegal. Oilseed production across SSA is projected to expand by an annual average of 2.3% p.a. to exceed 11 Mt by 2025, only 2% of global production.

In soybean production, South Africa stands in stark contrast to the rest of the region. Underpinned by a tripling of area and continuous yield improvements, soybean production expanded fivefold to exceed 1 Mt by 2015 from only 0.2 Mt between 2003 and 2005. Supported by rising demand from the animal feed sector, crushing capacity has expanded rapidly over the past few years, but soybean production failed to keep up. Thus South Africa moved from a traditional net exporter to a net importer from 2014. Sustained production growth of 7% p.a. is projected to 2025. As livestock production intensifies, crushing demand may also increase across the rest of the region. For instance in Zambia, soybean production is projected to expand by an annual average of more than 5% to 2025, albeit from a small base.

Other oilseed production is concentrated in the Western African region, with Nigeria alone producing 30% of the SSA total. Western Africa also accounts for more than half of projected production growth, however growth rates are more consistent across the region than for soybeans. The performance of selected countries illustrates that significant production growth is possible, but productivity remains very low by global standards leaving significant room for improvement. Meaningful expansion will also be dependent on the development and expansion of processing facilities.

SSA accounted for less than 2% of global protein meal consumption over the 2013-15 base period, reflecting the extensive nature of livestock production across most of the region. Protein meal use has expanded by more than 40% over the past decade, yet it remains concentrated in South Africa and Nigeria, which account for almost 60% of total use. As livestock sectors intensify in the coming years, protein meal use expands across most of SSA, with the fastest growth recorded in Western Africa (43%) and Eastern Africa (32%). In Southern Africa, projected growth is more modest at 16%, yet the base is much higher and in absolute volumes, Southern Africa accounts for the greatest share of additional protein meal use. The share of imports in total consumption declines marginally, mainly as a result of increased crushing volumes and reduced imports in South Africa.

Vegetable oil consumption in SSA has grown consistently over the past decade, yet at 11 kg per capita consumption remains well below the global average. Growth of 2.1% p.a. in per capita consumption makes it one of the fastest growing commodities in the region over the past decade. Growth is projected to be sustained, with Southern (1.4% p.a.) and Eastern Africa (1.2% p.a.) expanding the fastest to 2025. Given the limited oilseed processing facilities, imports comprise a substantial share of total consumption at more than 50% in Eastern and Southern Africa in the base period. High transportation rates therefore raise the cost of vegetable oil. Nonetheless, total imports into SSA are projected to expand by an annual average of 3.7%, of which Nigeria (4% p.a.), Sudan (5% p.a.), Ethiopia (6% p.a.) and Kenya (3% p.a.) account for the greatest share. Consequently the contribution of imports in total consumption in Eastern Africa increases further to almost 90% (Figure 2.19).

Figure 2.19. **Vegetable oil consumption in Sub-Saharan Africa**

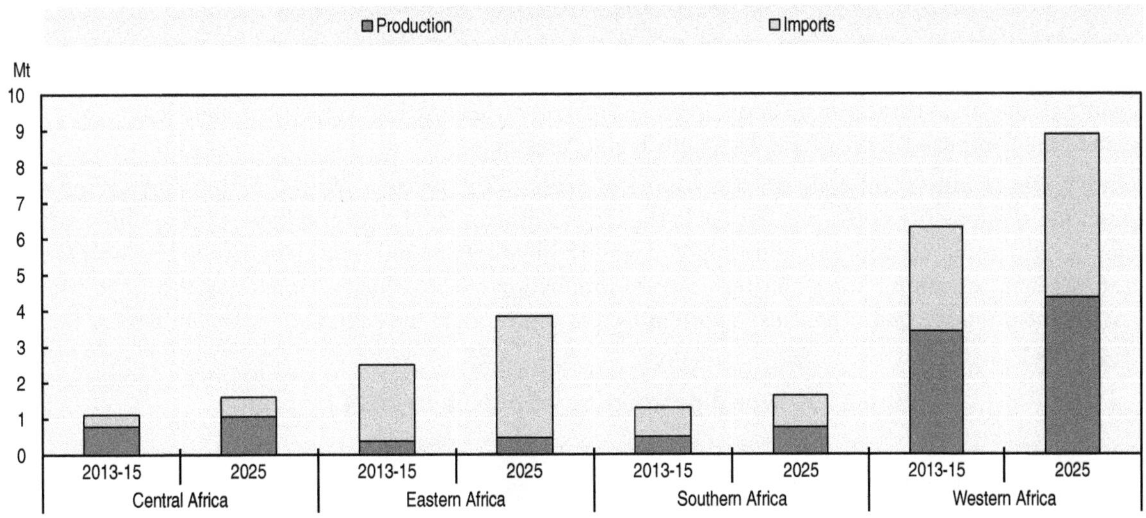

Source: OECD/FAO (2016), "OECD-FAO Agricultural Outlook", *OECD Agriculture statistics* (database), *http://dx.doi.org/10.1787/agr-data-en.*

StatLink http://dx.doi.org/10.1787/888933381520

Pulses

Pulses offer tremendous potential to alleviate malnutrition in SSA and its contribution to total protein intake is higher than in any other region in the world (Box 2.4). Per capita consumption remains well above the global average of 6.9 kg per capita in Central (10 kg), Eastern (22 kg) and Western Africa (17 kg), growth in per capita consumption of 2.5% p.a. (2.6 kg) over the past decade is sustained to 2025 and growth exceeds 2.5 kg per capita in all regions except Southern Africa, where consumption is already low in the base period.

Between 2013 and 2015, more than 50% of production originated in Eastern Africa, which also accounts for more than 65% of the additional 9.8 Mt produced across SSA by 2025. The popularity of pulses in SSA rests in their low input cost arising from the success of farm saved seeds, as well as their favourable impact on soil quality when planted alongside or in rotation with other crops such as maize.

Box 2.4. **2016 International year of pulses**

Pulses[1] have been an essential part of human nutrition for centuries and continue to be a major protein source and staple food in both developed and developing countries. Being dried seeds they can be stored for long periods without losing their nutritional value, allowing for flexibility and increased food availability between harvests. Crop residues can also be potentially used as feed and the heightened protein concentration from these are known to improve animal health.

Pulses are an increasingly important crop for smallholder farmers in a number of developing countries. The harvest can be consumed by the family or sold providing additional income. Being labour intensive, they provide employment opportunities both in the farm, during their cultivation, and off-farm through their processing. They have a lower carbon footprint than almost any other food group and have the ability of nitrogen fixation in crop rotation. Locally adapted pulses are drought-resistant and can be cultivated in very poor soils and semi-arid environments stabilising the food security situation in dry environments.

Box 2.4. **2016 International year of pulses** *(cont.)*

Canada is likely to continue being the world's leading exporter, shipping around 6 Mt of pulses (primarily dry peas) with a production of 5.8 Mt in 2015-16 expected to reach 7.2 Mt in 2016-17, followed by Australia, Myanmar, the United States and China. India is the world's leading importer, other significant importers include the European Union, China, Bangladesh and Pakistan.

India, where pulses are a significant source of protein for the poor as well as for vegetarians that constitute the majority of the population, is the largest consumer. The second largest being China which, together with India, accounts for almost half of world consumption. India has been the top producer for the past 30 years, accounting for a quarter of world production (20 Mt). Pulses production is a policy driven market and domestic grains policies in both India and China could have impact on the world market.

Figure 2.20. **World production of pulses in 2014 by region**

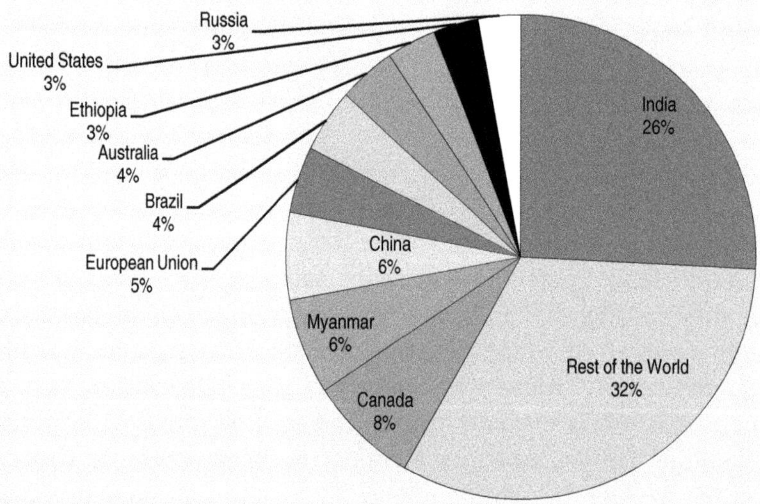

Source: FAOSTAT (2016). FAO, *http://faostat3.fao.org/*.

StatLink *http://dx.doi.org/10.1787/888933381538*

In general, consumption has seen a slow but steady decline. In addition to shifting diets in many countries, this may be partially due to an inability for production to keep pace with a growing population. Standard crop improvement methods complimented with modern biotechnology tools and genetic engineering are expected to play an important role in the generation of higher yields. The availability of innovations in developing countries will depend on continued significant levels of investments in agricultural research, both at the international and the national levels.

1. The term is limited to crops harvested solely for dry grain, thereby excluding crops harvested green for food (green peas, green beans, etc.) which are classified as vegetable crops. They include bambara beans, broad beans, chickpeas, cowpeas, dry beans, dry peas, lentils, lupines and vetches. For more detailed information on the International Year of Pulses please refer to the United Nations website *http://iyp2016.org/*.

Cotton

Cotton has emerged as an important cash crop in SSA, despite its small share in global production of 5.5%. Produced mainly for the export market, it has provided smallholder producers with a means to overcome input accessibility constraints through contract farming, playing a critical role in poverty alleviation in rural areas. The lower prices of man-made fibres, driven by substantially lower oil prices, have placed huge competitive pressures on world cotton markets in recent years. Despite potential opportunities for additional employment in downstream activities such as spinning and apparel, domestic use has declined over the past decade. Despite a partial recovery over the past five years, consumption remains well below the levels observed in the early 1990s and constituted less than 15% of production from 2013 to 2015. Consequently, cotton lint exports have gained increasing importance and the SSA share of global exports has grown to 15%. Given export orientated production, producer income remains sensitive to relative exchange rates as well as subsidised production in other areas of the world.

Production is concentrated in Western Africa, which accounts for more than 60% of the SSA total and where it is second only to cocoa beans in its contribution to agricultural export earnings. Despite a marginal decline in area, SSA production is projected to expand by 14%, surpassing 1.5 Mt by 2025. Production and export growth remains centred in four Western African countries that historically account for more than 55% of SSA production (Figure 2.21).

Despite revived domestic demand growth over the coming decade, 89% of SSA production will be exported by 2025. Export growth exceeds 1.7% p.a. in both Mali and Benin (Figure 2.21), though the sector continues to be challenged by infrastructural constraints; in particular in landlocked countries, where the time required to clear land borders creates bottlenecks that delay shipments. The value generated by cotton exports could therefore be increased if such challenges can be overcome.

Figure 2.21. **Cotton exports from selected countries in Sub-Saharan Africa**

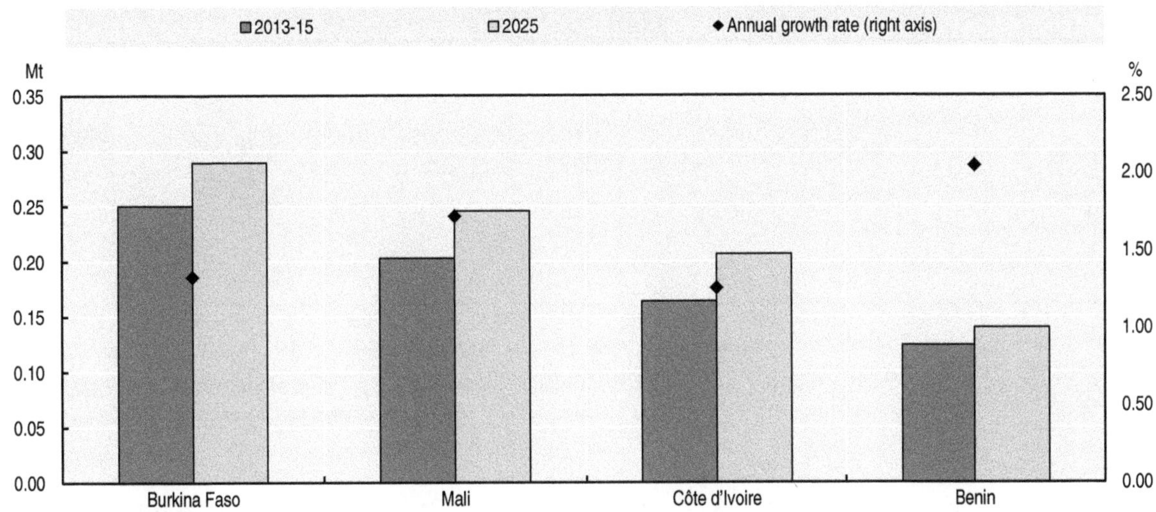

Source: OECD/FAO (2016), "OECD-FAO Agricultural Outlook", *OECD Agriculture statistics* (database), *http://dx.doi.org/10.1787/agr-data-en.*

StatLink http://dx.doi.org/10.1787/888933381548

Sugar

Sugar generally represents a success story within SSA agriculture. At an aggregate level, SSA is a net importer, yet several countries in Eastern and Southern Africa rank among the lower cost producers in the world and export consistently. Least Developed Countries in SSA have benefited from quotas providing preferential access to a lucrative sugar market in the European Union, supporting production growth over the past decade. Swaziland, Mauritius, Mozambique, Zambia, Malawi and Zimbabwe have all exported successfully to the European Union. Significant reforms to EU sugar policy however have presented an uncertain future for such exports. High transportation costs raise the cost of exports from landlocked countries and the expected reduction in EU prices following the reforms will likely shift exports away from the European Union into the SSA region.

Sugar consumption in SSA remains low in the global context and the average per capita consumption between 2013 and 2015 was only half the world average. Import demand from Western Africa in particular remains strong however (Figure 2.22) and is projected to expand by a further 34% over the next ten years relative to the 2013-15 base period. Consumption growth in Eastern and Southern Africa is also projected to expand by more than 2% p.a. and while import demand expands significantly in both Kenya and Tanzania towards 2025, production within Eastern and Southern Africa is sufficient for deficits to be met from neighbouring countries. Thus both regions retain a positive trade balance. In Eastern Africa in particular, raw sugar exports are complimented by refined sugar imports (Figure 2.22), suggesting that preferential access to the EU market has rendered raw sugar exports more profitable than domestic refinement.

Figure 2.22. **Net trade of sugar in Sub-Saharan Africa**

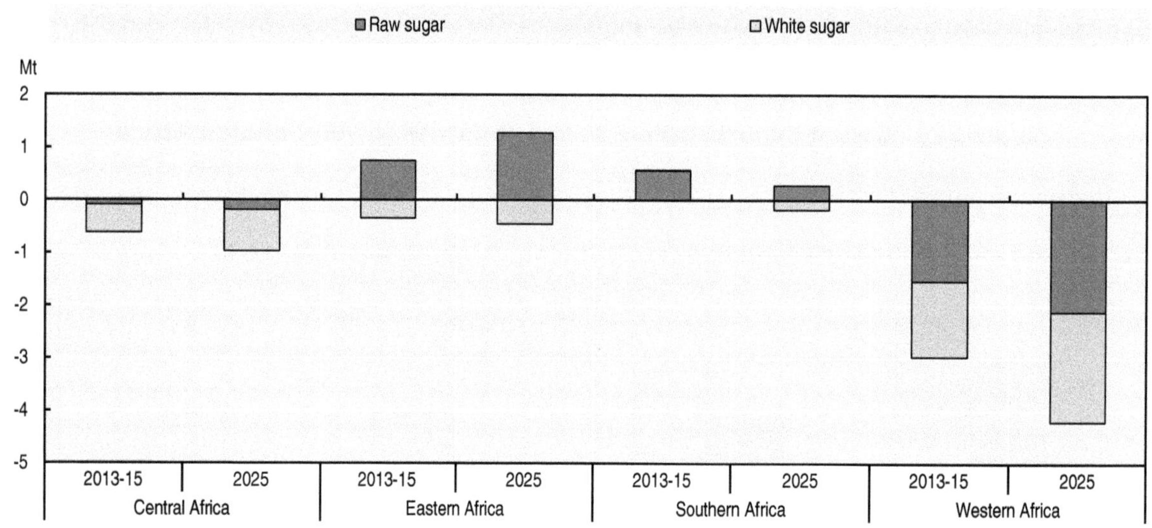

Source: OECD/FAO (2016), "OECD-FAO Agricultural Outlook", *OECD Agriculture statistics* (database), *http://dx.doi.org/10.1787/agr-data-en.*

StatLink *http://dx.doi.org/10.1787/888933381556*

Sugarcane production in SSA is projected to expand by approximately 1.7% p.a. over the coming decade, mainly from Eastern Africa, where Kenya and Mozambique register significant growth. Expansion is much slower in Southern Africa however, at only 0.2% p.a. for the ten-year period. South Africa is the largest sugar producer in SSA and its sugar area in excess of 300 kha is more than double that of any other country in the region. Not unlike many other countries, the sugar sector remains highly regulated. Despite its utilisation of a single channel marketing system, the South African sugar industry has struggled to come to terms with tighter profit margins, mainly due to stagnant and in some areas declining yields, combined with rising input and labour costs, and agrarian reform. Area under sugarcane has consolidated as a result and, exacerbated by drought conditions in 2015, exports have halved over the past decade. A substantial share of exports from Swaziland are already directed at the South African market, however reductions in exports to the European Union will likely increase the volumes flowing into South Africa from the rest of the region.

SSA possesses significant growth potential for sugar, but realisation of that potential will depend on the extent to which exports traditionally destined for the European Union can be absorbed within the region. Current import levels suggest that market space exists, yet trade diversion is hampered by excessive transportation rates. Opportunities for improved market access such as the EAC-COMESA-SADC tripartite free trade agreement will benefit surplus producers, though alternative domestic uses such as bioethanol production and cogeneration of electricity also provides possibilities for additional demand growth. Such ventures will however require a consistent regulatory framework related to their production.

Biofuels

The biofuel industry is small in SSA, comprising less than 1% of the global market. Several governments, however, have supported biofuel initiatives as a means of boosting economic growth and rural development. In many landlocked countries that rely on imported fossil fuels, biofuels have been promoted as a means towards increased energy security. Consequently many countries have introduced mandatory blending rates of varying levels. In addition some export oriented biofuel facilities have been built. Albeit from a small base, ethanol production has expanded by more than 90% over the past decade, with further growth of more than 3% p.a. projected to 2025. Production growth is concentrated in Southern and Western Africa, where it expands by 7% p.a.

Sugar provides the main feedstock for ethanol production, with some production also coming from sorghum. The use of food staple crops such as maize is uncommon and in many cases prohibited. Biodiesel production is a smaller industry than ethanol and over the past decade, production was only recorded in three countries across SSA – South Africa, Mozambique and Tanzania. Jatropha was long promoted as a crop with significant potential for biofuel production, yet it has not delivered the promised yields under adverse growing conditions and more recent studies have questioned its viability in semi-arid conditions due to high water requirements. Therefore the bulk of biodiesel expansion over the *Outlook* originates from South Africa, given the domestic biofuel program. Production is mainly from vegetable oil, but volumes remain small and, given that South Africa is still a net importer of vegetable oils, large scale expansion seems unlikely.

Meat and eggs

Per capita meat consumption in SSA at only 11 kg p.a. remains less than a third of the global average, yet significant regional differences are evident both in levels and composition (Figure 2.23). Meat consumption in Southern Africa is four times higher than any other region and while this is heavily influenced by South Africa, consumption in countries such as Namibia and Botswana is also well above the SSA average. Accounting for population however results in higher total meat consumption in Eastern and Western Africa, which together account for 54% of SSA meat consumption.

Notwithstanding the small base, the projected expansion of 35% in total meat consumption by 2025 outpaces any other region in the world. Underpinned by rising incomes, urbanisation and sustained population growth, robust consumption growth is projected across most of SSA, with an expansion of more than 38% evident in Central, Western and Eastern Africa. Slower growth of 20% in Southern Africa reflects a slowdown in South Africa, where per capita consumption has already surpassed 45 kg (Figure 2.23). Egg consumption provides an important alternative that reflects consumption growth of 36% over the ten year period. Consumption growth is also robust across the region and exceeds 50% in Eastern Africa.

Meat consumption preferences are somewhat unique in the region. Poultry accounts for 36% of total meat consumption between 2013 and 2015, but beef (33%) and sheep (19%) contributes a much greater share relative to the global average. This comes at the expense of pork consumption, which is significantly lower at 12%. Consumption preferences reflect cultural and religious preferences, as well as the dominance of extensive, pasture based production systems, with cattle grazing on communal pasture a common occurrence. Relative shares of different meats in the consumption basket remain fairly constant to 2025.

Figure 2.23. **Meat consumption in Sub-Saharan Africa**

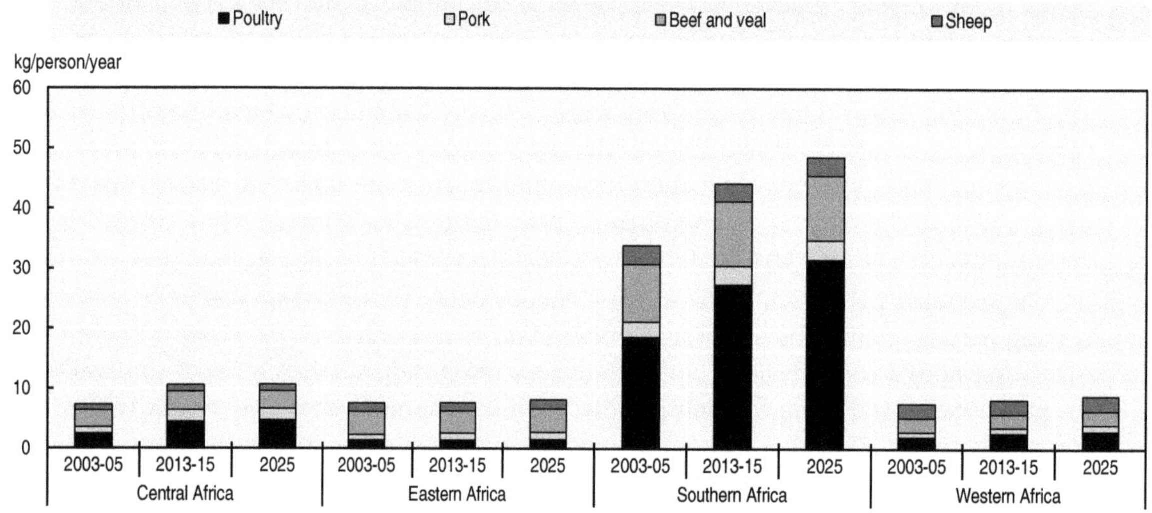

Source: OECD/FAO (2016), "OECD-FAO Agricultural Outlook", *OECD Agriculture statistics* (database), *http://dx.doi.org/10.1787/agr-data-en.*

StatLink *http://dx.doi.org/10.1787/888933381566*

Poultry consumption has expanded faster than consumption of any other meat in the past and with domestic supply unable to match demand, almost 40% of the additional consumption was imported. The preference for "dark" meat, which is less popular in many surplus production regions, has allowed imports to land at very competitive prices and volumes increased by an annual average of 13% over the past decade. Consumption growth in the largest importing countries such as South Africa and Angola is projected to slow, hence the rate of import growth reduces to 5% p.a. to 2025. South Africa is the biggest poultry producer in the region by some distance and is also responsible for the largest share of the 19% increase in production by 2025. This growth remains well below demand and more than half of the additional poultry consumed across SSA by 2025 will be imported (Figure 2.24).

Having expanded by close to 4% p.a. over the past decade, egg production in SSA increases by a further 750 kt by 2025, a rate of almost 3% p.a. Particularly strong growth is projected in Eastern (4% p.a.) and Western Africa (3% p.a.), which account for more than 70% of SSA egg production in 2025.

Figure 2.24. **Growth in meat demand in Sub-Saharan Africa**
2025 vs. 2013-15

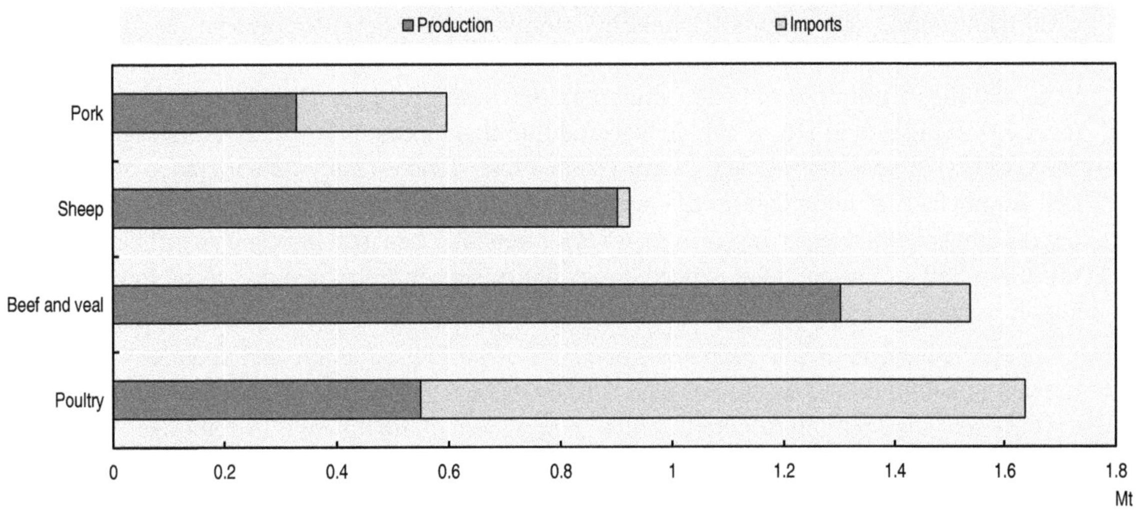

Source: OECD/FAO (2016), "OECD-FAO Agricultural Outlook", *OECD Agriculture statistics* (database), *http://dx.doi.org/10.1787/agr-data-en.*
StatLink *http://dx.doi.org/10.1787/888933381573*

Abundant grazing resources make extensive beef production an attractive prospect in SSA and in Southern Africa also an important export product, particularly given preferential access of Botswana and Namibia to the lucrative EU market. SSA is resident to 18% of the global bovine herd and consequently a great share of consumption is produced domestically. Growing herd numbers expanded beef production by almost 2% p.a. over the past decade despite the relative prevalence of disease outbreaks. Meat production per livestock unit remains well below the global average, suggesting that significant productivity gains could be achieved. Cattle also represent an important source of wealth in the region and many are kept for purposes other than beef production. Thus production per livestock unit is projected to rise only marginally in the coming decade and the bulk of the 27% increase in beef production is attributed to further herd expansion. Beef consumption growth is strong across the region, expanding by 2.6% p.a. to 2025. Growth is particularly strong in Eastern and Western Africa, where rates exceed 4% p.a. Within these regions, consumption growth is mainly driven by Kenya, Tanzania, Ethiopia, Zambia and Nigeria, all of which increase consumption by an annual average of at least 3%.

Sheep and pork consumption are much smaller in absolute terms, yet both industries grew impressively over the past decade. Growth in sheep meat consumption is sustained at a similar rate over the *Outlook*, while pork consumption slows from recent years. Pork consumption growth is robust across the region, but production does not respond sufficiently to meet demand and almost 45% of additional pork consumption will be imported. By contrast, sheep meat imports account for a very small share of additional consumption. Growth is significantly faster in Eastern Africa relative to the rest of SSA, where it is produced in an extensive, pasture based system. Eastern African production is centred in Sudan and Ethiopia, while Western African production is predominantly from Nigeria and Mali.

Fish

Fish and fishery products play an important role in food security in SSA, representing a valuable source of nutrients for healthy and diversified diets. Average per capita fish consumption in SSA is among the lowest in the world (8-9 kg vs. 19-20 kg of the world level), but the contribution of fish to animal protein intake is higher than the world average (over 20% compared to 17% at world level) and this share exceeds 50% in countries such as the Gambia, Ghana, Senegal and Sierra Leone. Projections reflect a 36% increase of food fish supply by 2025 compared to the average 2013-15 level, but accounting for significant population growth, the per capita increase is a mere 3%. Domestic supply is insufficient to meet demand and imports are expected to constitute an important share of the food fish supply, increasing by 32% in 2025 compared to the 2013-15 level.

Capture fisheries in SSA represents about 7% of world production, and about 40% of its harvest is within inland waters, of particular relevance in selected landlocked countries. Capture fisheries remain affected by the open access character of SSA fisheries. Many countries have focused more on production and revenue maximisation rather than sustainable management of resource productivity. This has caused over-exploitation of some valuable species, changes in the fish species composition and overall oscillation in catches by some countries in recent years. Together with weak monitoring capacity in many countries, these factors have increased the incidences of Illegal Unregulated and Unreported (IUU) fishing activities. The role of foreign fishing vessels, which fish in several coastal waters of the region, often under foreign access agreement, is also noteworthy, as it has reduced benefits for resource-adjacent countries, due largely to poor or weak negotiations of terms of agreement. The increasing adoption of more conservative management measures by some countries, including improved access control (e.g. registration, licensing systems) are expected to increase total capture fisheries production by 15% by the end of next decade compared to the 2013-15 average level.

Aquaculture has been introduced to most countries in SSA, though the region currently produces only 1% of the world's farmed fish. In the last few decades, its slow pace has frustrated the attempts of internal development agencies, governments and private sector investors. Yet in the long run, the enormous potential of aquaculture is still widely acknowledged as important to overcoming food security and nutrition challenges of the region. While challenges remain, prospects appear to be improving in a number of countries. Coupled with the spread of improved farming techniques and facilities, growth has been encouraging in many locations. Appropriate policies by some governments have allowed the private sector to lead aquaculture development, resulting in the emergence and intensification of small-scale and medium-size enterprises, market-led and large-scale commercial initiatives. More substantial regional growth is expected in the coming years, with overall production growing by 84% in 2025 compared to the average 2013-15 level. Aquaculture has already increased its share in total fisheries production in SSA from only 1% in 2004 to 8% in 2014. It is estimated that this share will be about 12% in 2025.

Dairy

Milk production's enormous potential in economic development and food security in rural areas makes dairy an important subsector in SSA. Particularly in Southern and Eastern Africa, commercialisation of the sector has illustrated dairy's potential to provide a regular income source that reduces poverty and improves living standards. Eastern Africa currently constitutes more than half of total milk production in SSA and a vibrant smallholder farming sector made a considerable contribution to milk production growth of 37% over the past decade. Sustained production growth is projected for the Outlook, rising by an annual average of 2.7% in Eastern Africa and 2.5% in SSA. Kenya's dairy sector represents a particularly well developed value chain in the region, with a range of small, medium and large scale producers and accounts for almost 15% of the additional milk production. Support services in the sector are more developed than other parts of SSA and underpin the success of the sector.

Dairy represents a primary protein source to SSA consumers and fresh dairy products account for more than 90% of total dairy consumption. Demand for dairy products expanded by 1.8% p.a. over the past decade and in light of continued income growth and urbanisation; growth is projected to accelerate to 2.6% p.a. over projection period. Consumption in Eastern Africa is significantly higher than the rest of SSA (Figure 2.25), supported by per capita consumption in excess of 100 kg in Somalia, Sudan and Kenya. Such levels are not only significantly higher than the rest of the region, but also well above the global average. Nonetheless, projected growth is strong at 2.6% p.a. over the coming decade, although consumption remains low in several countries.

Figure 2.25. **Dairy product consumption in Sub-Saharan Africa**

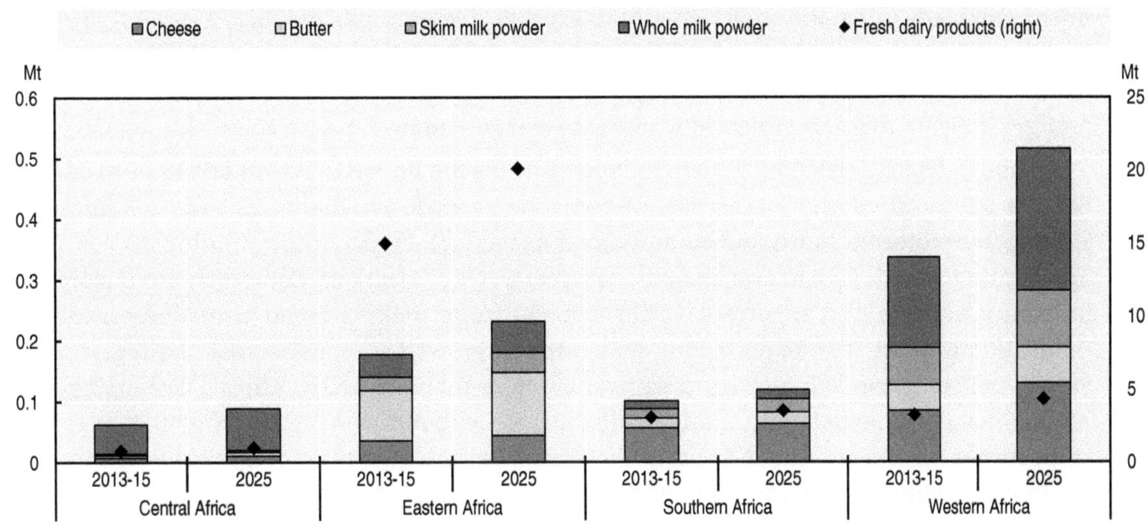

Source: OECD/FAO (2016), "OECD-FAO Agricultural Outlook", *OECD Agriculture statistics* (database), *http://dx.doi.org/10.1787/agr-data-en*.

StatLink http://dx.doi.org/10.1787/888933381581

Given the high share of fresh products in consumption, trade accounts for a very small share of the domestic market. However in Western Africa, skim and whole milk powder account for almost 10% of total dairy consumption – the bulk of which is imported. This trend is projected to continue and supported by strong demand. Skim and whole milk powder imports into Western Africa are projected to expand by more than 3% p.a. to exceed 300 kt by 2025. Although not presently included in modelled commodities, fat filled milk powder is another important dairy product that constitutes a significant share of dairy product imports into Central and Western Africa in particular.

Fruit and beverage crops

A number of developing countries in SSA rely heavily on the export of primary commodities, such as tropical beverages crops, fruits, and raw materials for the bulk of their export revenues. Such exports constitute an important source of revenue for smallholder producers and provide rural households with employment opportunities at the farm level and throughout the value chain. At the macro-economic level, the production and export of tropical beverages crops, fruits, and raw materials support government fiscal resources through tax revenues, and contribute to foreign currency reserves that facilitate imports of food, and other goods and services. Beyond these benefits, consumption has important nutritional benefits that support food security and sales proceeds enable access to essential staples as well as other goods and services.

The commodities responsible for generating export revenues differ across regions, with wine and fruits such as citrus and table grapes providing the bulk of agricultural export revenue in Southern Africa, compared to cocoa beans and bananas in Central and Western Africa. In East Africa, tobacco is important, as well as beverage crops such as coffee and tea.

South Africa is the main fruit exporter in the Southern African region and historically, citrus leads the way. South Africa is the second largest citrus exporter in the world and the largest in the southern hemisphere. Production is projected to expand continuously and given its high share of total citrus production, oranges account for the largest share of additional production. Relative price shifts induce rapid growth in soft citrus, lemon and lime production. Given the predominance of exports, domestic processing is a small share of the market. Thus orange juice production expands by an annual average of only 0.5%, compared to projected growth in orange exports of 3.5% p.a. Domestic demand for fresh oranges in South Africa is projected to increase by just under 2% p.a. to 2025.

Though typically smaller industries, tropical fruits are important crops across most of SSA. In South Africa, more than 50% of domestic avocado production is exported and exports are projected to expand by an annual average of 4% p.a. in the coming decade, accelerating from 3.4% p.a. over the past ten years. Ivory Coast and Cameroon are major banana producers and exporters. Other tropical fruits make a larger contribution to domestic nutrition than exports, but consumption growth is robust across the region. Papaya consumption is projected to expand by more than 3% p.a. in Africa's developing regions, while pineapple and mango consumption expand by 4.5% p.a. and 4.1% p.a. respectively. Nigeria and Sudan in particular consume large quantities of tropical fruit and the bulk of additional demand is met through domestic production.

Beverage crops represent a very important component of agricultural export revenue, particularly in Eastern Africa. Kenya is the largest tea exporter by some distance and production growth of 3.8% p.a. in the coming decade is sufficient to support export growth of 3.2% p.a., despite firm growth in domestic consumption. Robust export growth in excess of 5% p.a. is also projected in Rwanda and Uganda, albeit from a significantly smaller base.

Challenges and uncertainties

The outlook for agriculture in Sub-Saharan Africa provides many reasons for optimism, but there are major challenges and uncertainties. Demand growth is supported by a rapidly expanding population, combined with firm projected income growth. However income levels across large parts of the region remain very low and growth is from a small base. Maintaining and improving the political and economic conditions for agricultural production growth and food security advances will be crucial. For the vast majority of the region, there is only limited fiscal space to counter a slowdown and growth may depend on the extent to which governments are able to anticipate shocks to the system. Implementation of policies that maintain macro-economic stability is essential for local and foreign direct investments to continue. For resource-rich countries whose growth has been supported largely by the commodity boom, the extent of investment into economic diversification and integration into global value chains will shape future growth trajectories under a cycle of lower oil and other commodity prices.

From a demand perspective, the distribution of income growth in the coming decade may be as important as the rate. Despite impressive economic growth in the past, poverty rates remain stubbornly high and women, who are a crucial resource in agriculture and the rural economy, are disproportionately affected. Not only are they over represented in unpaid, seasonal and part-time work but they face a number of constraints in accessing agricultural inputs, services and markets that inhibit their pathways out of poverty (FAO, 2015). Greater concentration of consumer demand and land ownership may restrict the breadth of economic growth and slow the poverty-reducing benefits of whatever agricultural growth does occur. Most crucially, income growth restricted to a narrow segment of society would diminish the income multiplier effects that otherwise might encourage more rapid and inclusive economic transformation. By contrast, broad based income growth has the potential to meaningfully reduce poverty, resulting in a rising number of middle to higher income consumers. Large numbers of consumers with the financial means for greater dietary diversity will not only increase total demand for food products significantly, but also change the composition of demand, shifting away from traditional starches toward animal products, fresh fruit and vegetables, as well as more convenient, processed foods. Per capita meat consumption in particular remains very low in the region and significant changes to existing dietary preferences could result in a vastly different outlook on the demand side.

From a supply perspective, one of the greatest challenges faced by the region relates to the slow rate of productivity gains. In this regard, an important uncertainty that will have a far ranging impact on production practices and productivity growth is the extent of concentration of agricultural land, which in turn will also be influenced by land tenure policies. Increasing concentration and commercialisation of medium-scale farmers could accelerate the rate of technology adoption, which has been fairly slow to date. Efficiency gains by a growing number of small, medium and large-scale farms linked into vertically integrated value chain with greater opportunity for access to credit, technology, extension services and off-take agreements, could have a meaningful impact on output levels in the coming decade. Commensurate development of upstream and downstream food sectors could increase the opportunities for non-farm income, which may in turn provide relatively productive small-scale producers with the capital to break through the barriers of subsistence agriculture into more commercialised medium-scale stature.

Productivity also remains low in livestock production and intensive production of pork and poultry has only taken off in selected countries. The region remains a net importer of most feed grains and protein meal, resulting in high prices that are not conducive to investment in intensive production systems. Increased productivity that induces surplus feed grain production accompanied by lower prices could however induce such investment, causing significantly higher feed demand.

Considering the severity of the impact of the 2015-16 drought on food security in the region (Box 2.3), the potential impact of climate change cannot be ignored. The frequency of drought occurrence is already higher in SSA relative to most other regions in the world and agricultural production remains largely rain-fed. While the precise impacts of climate change on African farming systems are likely to vary spatially in ways that are difficult to predict, two general predictions for which there is now some consensus are greater variability in agricultural production and a possible decline in crop productivity (Schlenker and Lobell, 2010). The evolution of both farm structures and farming practices in the region will impact on the resilience to increasing climatic variability in the future. Increasing the rate of technological adoption, facilitating access to irrigation systems and improved farming practices that support such resilience remains one of the greatest challenges facing the region.

Arguably the greatest challenge facing the agricultural sector in SSA is weak infrastructure including transportation networks, access to energy, irrigation systems and stockholding facilities. Poor transportation networks limit access to markets, often exacerbate high levels of post-harvest losses and also inhibit efficient distribution of inputs such as seed and fertiliser. At the same time, it is an underlying factor in high food prices, as it raises the cost of both inputs and imported food products. Substantial differences in price levels between surplus and deficit regions suggest that investments able to reduce the cost of transportation would hold significant benefits to producers and consumers alike. Not only would it reduce the cost of imported food products to bolster demand, but it would also provide a more lucrative export market for surplus producers.

Conclusions

The significance of the agricultural sector in SSA is reflected in its high share of GDP in most countries, its even greater share of employment and its prioritisation in the development agenda. While the total value of agricultural output has grown markedly over the past decade, SSA remains the most food insecure region in the world, with uneven progress towards eradicating hunger over the past decade. The Malabo Declaration on accelerated agricultural growth strives to eradicate hunger in Africa by 2025. Among other objectives, it targets a doubling of agricultural productivity, a halving of current levels of post-harvest losses and a threefold increase in intra-regional trade levels. Within the context of resilient agricultural systems, it also targets social protection systems and decent employment opportunities for rural populations.

In light of agriculture's clear role in confronting the challenge of eradicating hunger and improving food security, this chapter provides an outlook for agriculture in SSA that considers the complexities associated with the region. The development of the sector in the coming decade will continue to be shaped by policies and megatrends which remain subject to high degrees of uncertainty. These megatrends include factors that will shape food demand such as rapid population growth, income growth (with uncertainties over the rate and distribution), the consequent rise of an African middle class, rapid urbanisation, facilitated access to new information technologies and communication combined with continued expansion of rural population numbers and agriculture's likely role as the single largest source of employment to multitudes of young people entering the labour force.

In light of such factors, the *Outlook* portrays optimistic growth in food demand of more than 3% p.a. towards 2025, yet much of that increase remains driven by population growth and increases in per capita intake remain modest across most of the region. SSA exhibits exceptional diversity in income levels and consumption preferences, hence the product mix also exhibits regional variation, but robust growth in caloric intake derived from vegetable oil and sugar is evident across most of the SSA region. Increasing diversity in the production mix also provides opportunities to improve dietary diversity and increase potential income generation. The prevalence of undernourishment has been reduced to 5.2% in Southern Africa in 2014-16, yet in Central Africa it remains above 40%. Given rapid population growth, SSA will account for a rising share of the global total of undernourished (Box 1.3). Commensurate with income growth, the greatest increase in per capita caloric and protein intake over the outlook period is projected in East Africa, while the slowest growth rate is attributed to Central Africa, which remains challenged by political instability and civil strife.

SSA exhibits vast agricultural potential, yet production growth in the past has mainly been achieved through continued area expansion. While total agricultural production is projected to rise by 2.6% p.a. to 2025, area expansion slows and an increasing share of production growth is attributed to improved productivity. Multiple factors influence accelerated productivity gains over the outlook, including faster technology adoption associated with the emergence of medium-scale producers and improved integration of smallholder producers into the value chain. Despite improvements, significant yield gaps remain and imports of most primary food products are projected to rise. The extent to which current yield gaps can be closed represents one of the greatest challenges and uncertainties facing the region.

The potential contribution of the agricultural sector to poverty reduction, improved livelihoods of rural households and greater food security in SSA is undisputed and the outlook presented in this chapter remains broadly positive. Yet growth in the sector remains challenged by an uncertain policy environment and poor infrastructural development that limit market access, increase post-harvest losses and raise the cost of trade. Epizootic and climatic events also challenge the medium-term development of the agricultural sector. Thus food prices in the region remain high, which impacts negatively on food security, particularly given that most small scale producers are still net buyers of food products. Significant price differences remain across the region and increased intra-regional trade offers opportunities to improve food security and reduce poverty.

Strategic investment by both public and private sector has the ability to further improve the outlook presented in this chapter. Abundant interest from both foreign and domestic investors has at times been hampered by inconsistent policy application. Thus while public investments into infrastructure, research and extension is critical, the institution of an enabling environment that promotes private investment and job creation in both farming and non-farm sectors will have high pay-offs that are able to smooth continued economic transformation in a region with undoubted potential. Effective implementation of investment strategies at national and continental level will aid in achieving hunger eradication targets, and in transforming food systems in Africa for inclusive growth and shared prosperity.

Notes

1. The Sub-Saharan African region is defined by the United Nations Statistical Division and is used to indicate all of Africa, except Northern Africa, with Sudan included in Sub-Saharan Africa. Regional aggregations are available at *http://unstats.un.org/unsd/methods/m49/m49regin.htm* and detailed in the glossary.
2. A megatrend is a social, economic, political, environmental, or technological change that is typically slow to form yet, when in place, exerts major influence on human behaviour (Jayne et al., 2014). Most of the megatrends mentioned here are detailed in Jayne et al., 2014.
3. The 2014 *African Economic Outlook* report projects that foreign investment and official remittances to Africa could reach more than USD 80.0 billion and USD 67.1 billion, respectively, in 2014.
4. Fuglie and Rada (2013) report that fallowed land as a proportion of total farmland in SSA has declined from 40% in 1960 to roughly 15% in 2011. Jayne et al. (2014b) report that fallows have largely been eliminated in smallholder farming areas containing more than 250 people per km^2 of arable land.
5. Full details on the Malabo Declaration are available from *http://pages.au.int/sites/default/files/Malabo%20Declaration%202014_11%2026-.pdf.*
6. The countries included in the MAFAP study are: Burkina Faso, Ethiopia, Ghana, Kenya, Malawi, Mali, Mozambique, Nigeria, Uganda and the United Republic of Tanzania.

References

AfDB (African Development Bank) (2011), "The Middle of the Pyramid: Dynamics of the Middle Class in Africa", *Market Brief*, April, *www.afdb.org/fileadmin/uploads/afdb/Documents/Publications/The%20Middle%20of%20the%20Pyramid_The%20Middle%20of%20the%20Pyramid.pdf.*

AfDB, FAO and ECOWAS (African Development Bank, Food and Agriculture Organization of the United Nations, Economic Community of West African States) (2015), "Trade Policy" in *Agricultural Growth in West Africa: Market and policy drivers*, Food and Agriculture Organization Publications, Rome, *www.fao.org/3/a-i4337e.pdf.*

AfDB, OECD and UNDP (2014), African Economic Outlook 2014: Global Value Chains and Africa's Industrialisation, OECD Publishing, Paris, *http://dx.doi.org/10.1787/aeo-2014-en.*

Allen, T. and P. Heinrigs (2016), "Emerging Opportunities in the West African Food Economy", *OECD West African Papers*, OECD Publishing, Paris

Alliance for a Green Revolution in Africa (AGRA) (2014), Africa Agriculture Status Report 2014: Climate Change and Smallholder Agriculture in Sub Saharan Africa, *http://hdl.handle.net/10568/42343.*

Angelucci F., J. Balié, H. Gourichon, A. Mas Aparisi, M. Witwer (2013), "Monitoring and Analysing Food and Agricultural Policies in Africa. MAFAP Synthesis Report 2013", MAFAP Synthesis Report Series, Food and Agriculture Organization Publications, Rome.

Brink. A. B. and H.D. Eva (2009), "Monitoring 25 years of land cover change dynamics in Africa: A sample based remote sensing approach, Applied Geography, Volume 29, Issue 4, December, pp. 501-512, *http://0-dx.doi.org.innopac.up.ac.za/10.1016/j.apgeog.2008.10.004.*

Chamberlin, J., T.S. Jayne, and D. Headey (2014), "Scarcity amid Abundance? Reassessing the Potential for Cropland Expansion in Africa." Food Policy, Vol. 48: 51-65.

Chapoto, A and N. Sitko (2014), "Understanding the effects of trade restrictions on maize prices", Presentation to COMESA/ACTESA Parliamentary Policy Seminar on Import and Export Bans, Lusaka, 11-12 August. http://fsg.afre.msu.edu/zambia/COMESA_PRESENTATION.pdf.

Drechsel, P., L. Gyiele, D. Kunze, and O. Cofie (2001), "Population Density, Soil Nutrient Depletion, and Economic Growth in Sub-Saharan Africa", Ecological Economics, Vol. 38, pp. 251-258.

FAO (2016, forthcoming), Medium-term Outlook for Raw Materials, Horticulture and Tropical, Food and Agriculture Organization Publications, Rome.

FAO, IFAD and WFP (2015), The State of Food Insecurity in the World 2015. Meeting the 2015 international hunger targets: taking stock of uneven progress, Food and Agriculture Organization Publications, Rome.

FAO, CTA and IFAD (2014), Youth and Agriculture: Key challenges and concrete solutions, Food and Agriculture Organization Publications, Rome.

Filmer, D., and L. Fox (2014), Youth Employment in Sub-Saharan Africa, Africa Development Series, World Bank, Washington, DC. Doi:10.1596/978-1-4648-0107-5.

Fine, D., A. van Wamelen, S. Lund, A. Cabral, M. Taoufiki, N. Dörr, A. Leke, C. Roxburgh, J. Schubert, and P. Cook (2012), Africa at Work: Job Creation and Inclusive Growth, McKinsey Global Institute, Boston.

Fuglie, K. and N. Rada (2013), "Resources, policies, and agricultural productivity in sub-Saharan Africa", Economic Research Report, N°145, US Department of Agriculture Economic Research Service, Washington, DC.

Global Information and Early Warning System on Food and Agriculture (GIEWS) (2016), "Delayed onset of seasonal rains in parts of Southern Africa raises serious concern for crop and livestock production in 2016", GIEWS Special Alert No. 336, Food and Agriculture Organization of the United Nations (FAO), Rome. [Online] Available at: *www.fao.org/3/a-I5258e.pdf.*

IMF (International Monetary Fund) (2012), International Jobs Report, Economist Intelligence Unit, Washington, DC.

IMF (International Monetary Fund) (2015), International Jobs Report, Economist Intelligence Unit, Washington, DC.

Jayne, T.S., K. Yeboah, J. Chamberlin, L. Traub, M. Muyanga, N. Sitko, A. Chapoto, C. Nkonde, W. Anseeuw, and R. Kachule (2015), "Africa's Changing Farmland Ownership: Causes and Consequences", Plenary paper presented at the 29th International Conference of Agricultural Economists, Milan, Italy, August.

Jayne, T.S., F.H. Meyer, and L. Traub (2014), "Africa's Evolving Food Systems: Drivers of change and the scope for influencing them", IIED Working Paper, IIED, London.

Jayne, T.S., D. Headey, and J. Chamberlin (2014), "Land Pressures, the Evolution of Farming Systems, and Development Strategies in Africa: A Synthesis", Food Policy, Vol. 48, pp. 1-17.

Jayne T.S. and S. Rashid (2013), "Input Subsidy Programs in Sub-Saharan Africa: A Synthesis of Recent Evidence." Agricultural Economics, Vol. 44 (6), pp. 1-16.

Jayne, T.S. and D.L. Tschirley (2009), "Food price spikes and strategic interactions between the public and private sectors: Market failures or governance failures", Commodity Market Review 2010, pp. 3-18.

Losch, B. (2012), "Agriculture: the key to the employment challenge", Perspective No. 19, CIRAD, Montpellier.

Minot, N. (2014), "Food price volatility in sub-Saharan Africa: Has it really increased?" Food Policy, Vol. 45, pp. 45-56.

Morrison, J. and A. Sarris (2016), *Food Staple Market Volatility and Food Security in Eastern and Southern Africa: What role for trade and market policy?*, Food and Agriculture Organization Publications, Rome.

New Partnership for Africa's Development (NEPAD) (2014), *Agriculture in Africa: Transformation and Outlook*, Online available at: *www.nepad.org/system/files/Agriculture%20in%20Africa.pdf*

OECD (2013), *Settlement, Market and Food Security*, West African Studies, OECD Publishing, Paris, *http://dx.doi.org/10.1787/9789264187443-en.*

ReNAPRI (2015), "Anticipating the Future of Agriculture in the Region: Outlook for Maize, Wheat, Sugar and Rice", Lusaka, Zambia.

Schlenker, D., and D. Lobell (2010), "Robust Negative Impacts of Climate Change on African Agriculture." *Environmental Research Letters*, Vol. 5 (1).

Stoorvogel, J.J., and E. Smaling (1990), "Assessment of Soil Nutrient Depletion in Sub-Saharan Africa: 1983-2000", *Nutrient Balances per Crop and per Land Use System*, Report No. 28, Vol. 2, Winand Staring Center, Wageningen.

Tittonell, P., and K. Giller (2012), "When Yield Gaps Are Poverty Traps: The Paradigm of Ecological Intensification in African Smallholder Agriculture", *Field Crops Research*, Vol. 143 (1), pp. 76-90.

Yeboah, K., and T. S. Jayne (2015, forthcoming), "Employment Trends in Sub-Saharan Africa: The Evolving Role of Agriculture", *International Development Working Paper*, Michigan State University, East Lansing.

World Bank (2016), *World Development Indicators* [Online consultation 15 March 2016] Available at: *http://data.worldbank.org/data-catalog/world-development-indicators.*

OECD-FAO Agricultural Outlook 2016-2025

PART I

Chapter 3

Commodity snapshots

This chapter describes the market situation and highlights of the latest set of quantitative medium-term projections for world and national agricultural markets, for the ten-year period 2016-25. It provides information on prices, production, consumption, trade and main uncertainties for cereals, oilseeds, sugar, meat, dairy products, fish, biofuels and cotton. The quantitative projections are developed with the aid of the partial equilibrium Aglink-Cosimo model of world agriculture. The printed version of this chapter only includes the projection highlights for each commodity whereas further details and an extensive statistical annex are available on line.

CEREALS

Market situation

Global cereal markets have been characterised over the past few years by abundant supplies amid slower demand growth. As a result, world inventories have increased and international prices of all cereals have fallen to relatively low levels compared to the previous decade. Even the decline in world cereal production in 2015, following the 2014 record harvest, could not reverse this downward pressure, leading to further declines in international prices during the 2015 marketing year (see glossary for a definition of marketing year). Given the early prospects in world cereal output for this season, weak demand and large inventories in 2016, global markets are likely to experience relatively low prices. Against this background, only radical or sudden changes in demand or supply are likely to alter the short-term outlook.

Projection highlights

Starting with relatively low prices during the base period (2013-15), the prevailing sluggish economic growth conditions, large stocks, low oil prices and a strong US dollar are likely to keep prices under pressure in the short run. Over the course of the medium-term, however, prices of wheat and coarse grains (in nominal terms) are projected to be primarily cost driven, increasing in nominal terms but not by enough to keep pace with inflation, implying slight declines in real terms. However, prices of all cereals even in nominal terms are projected to be lower on average than in the previous decade, although well above the levels of the decade before.

Global cereal production is projected to expand by 12% by 2025 from the base period, mainly driven by yield improvements, with limited area expansion. Compared with the base period, production of wheat in 2025 is projected to increase by 10% (71 Mt), with India producing 10 Mt more, the People's Republic of China (hereafter "China") 7.9 Mt Argentina 5.6 Mt, Ukraine 5 Mt, the Islamic Republic of Iran 4.7 Mt, Turkey 4.2 Mt, the European Union 3.5 Mt, the Russian Federation 3.1 Mt, Canada 1.9 Mt[1] and the United States 1.7 Mt. Rice production is set to increase by 14% (69 Mt), with most of the increase (59 Mt) concentrated in Asian countries, led by India (20 Mt), Indonesia (8.1 Mt), Viet Nam (6 Mt), Bangladesh and China (4 Mt each), as well as Thailand (2.8 Mt). Maize production is projected to rise by 13% (131 Mt), led by the United States (27 Mt), Brazil (21.5 Mt), China (21 Mt), Argentina (6 Mt), the European Union (5.6 Mt), and Indonesia (4 Mt). Production of other coarse grains is projected to increase by 8%, or 25 Mt, with the biggest increases in Ethiopia (5.5 Mt), Argentina (3.1 Mt) and India (2.9 Mt), followed closely by Nigeria (1.9 Mt).

Global cereal use is projected to grow by 14% or 340 Mt, to reach 2 818 Mt by 2025. Wheat consumption is expected to increase by 11% compared to the base period and continues to be largely used for human consumption (69% of total use throughout the projection period). The use of wheat for feed is projected to increase, mostly in China, the Russian Federation and the European Union, while biofuel use of wheat only accounts for 1.2% of global use in 2025. Maize use for animal feed is projected to increase its overall share over total use from 56% in the base period to 60% in 2025. The projected increase in total maize utilisation (157 Mt), is mainly driven by higher feed use (127 Mt) – mostly on account of fast expanding livestock sectors in developing countries. Maize for human consumption is projected to grow by 21% (28 Mt), mainly in developing countries, especially those in Africa where white maize is a main staple in several countries. The use of other coarse grains is also set to grow, by 11% (31 Mt), driven mainly by food demand (16 Mt) followed closely by feed demand (14 Mt). The expansion of food use mainly comes from Sub-Saharan Africa (13 Mt), while China accounts for most of the expansion for feed. Direct human consumption remains the main end-use of rice, as a major staple food in large parts of Asia, Africa, Latin America and the Caribbean. Total consumption is predicted to rise to 563 Mt by 2025, sustained principally by population growth. Given the expected demographic changes, Asian countries are anticipated to account for more than 80% of the projected increase in global rice consumption.

Figure 3.1. **World cereal prices**

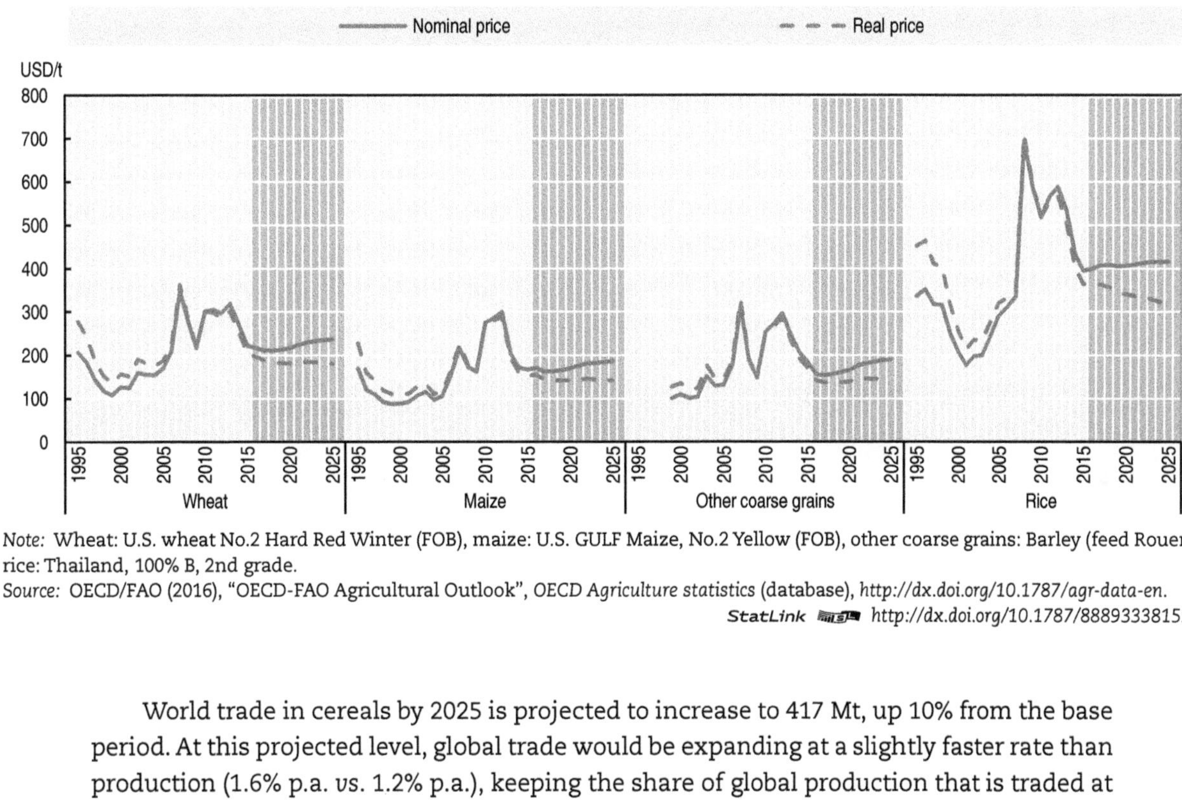

Note: Wheat: U.S. wheat No.2 Hard Red Winter (FOB), maize: U.S. GULF Maize, No.2 Yellow (FOB), other coarse grains: Barley (feed Rouen), rice: Thailand, 100% B, 2nd grade.

Source: OECD/FAO (2016), "OECD-FAO Agricultural Outlook", *OECD Agriculture statistics* (database), *http://dx.doi.org/10.1787/agr-data-en.*

StatLink http://dx.doi.org/10.1787/888933381591

World trade in cereals by 2025 is projected to increase to 417 Mt, up 10% from the base period. At this projected level, global trade would be expanding at a slightly faster rate than production (1.6% p.a. *vs.* 1.2% p.a.), keeping the share of global production that is traded at 15%. For wheat, this share is expected to reach 22% by 2025, compared with 12% for maize and 15% for other coarse grains. Continuing historical trends, developed countries are expected to remain as the main exporters of wheat and coarse grains to developing countries, while rice is mostly traded between developing countries. The global players on international rice markets are expected to remain consistent, although exporters such as Cambodia and Myanmar are projected to increase their shares of the international market over the decade.

The anticipated continuation of lower cereal prices compared to the previous decade will impact on planting decisions and hence supply responses. Relative prices to other crops like oilseeds are therefore an important factor over the next decade which might lead to stronger reallocation of crops. On the demand side, developments in the fastest growing economies will have more profound implications for trade. Demand changes in China and their timing of releasing maize stocks are main uncertainties during the projection period.

The expanded cereals chapter is available at
http://dx.doi.org/10.1787/agr_outlook-2016-7-en

OILSEEDS AND OILSEED PRODUCTS

Market situation

Global soybean production for the 2015 marketing year (see glossary for a definition of marketing year) continued to increase, whereas production of other oilseeds (rapeseed, sunflower seed and groundnuts) declined relative to 2014. Low crude oil and cereal prices put additional pressure on oilseed prices.

Vegetable oil production increased more slowly than oilseed production for two reasons. First, palm oil yields decreased in Southeast Asia due to *El Niño* and, second, the slow production growth of oilseed oils due to an increased share of soybeans (containing less oil than other oilseeds) in the oilseeds market. However, growth in vegetable oil demand has slowed recently due to contracting biodiesel production from vegetable oils in 2015 in several developed and developing countries. Vegetable oil prices are expected to recover first within the oilseed complex due to currently stagnating production.

The continuously growing demand for protein meals has been the main driver behind the expansion of oilseed production in recent years. This has increased the share of protein meals in the returns from the crushing of oilseeds, and more so for soybeans over other oilseeds due to its higher protein content. Compared with coarse grains and other feed ingredients, protein meal prices have declined recently to historically average levels, meaning that protein meal prices are about 1.5 to 2 times those of maize.

Projection highlights

In nominal terms all oilseeds and oilseed product prices are projected to increase over the outlook period. The price relationships within the sector will shift slightly in favour of the meal component. Due to saturation in per capita food demand in many emerging economies and reduced growth in biodiesel production from vegetable oils, vegetable oil prices will decline whereas protein meal prices will increase slightly in real terms during the outlook period.

During the outlook period, global soybean production is expected to continue its expansion, yet at 2.4%, below the annual growth rate of 4.2% experienced during the last decade. Production of other oilseeds increases by 1.2% p.a. over the next decade, considerably below the growth rate of 3.6% p.a. in the previous decade. Globally, crushing soybean and other oilseeds into meal (cake) and oil dominates total usage and it increases slightly faster than other uses, notably direct food consumption of soybeans, groundnuts and sunflower seed. Overall, 91% of world soybean production and 84% of world production of other oilseeds will be crushed in 2025.

Vegetable oil includes oil from the crushing of soybeans and other oilseeds (around 55% of production), palm (36%), as well as palm kernel, coconut and cottonseed oils. World vegetable oil production will remain concentrated among a few countries in the coming decade. Despite a slowdown in area expansion, significant production growth still occurs in the main palm oil producing countries: Indonesia (2.5% p.a. vs. 8.1% p.a. in the previous decade) and Malaysia (2.1% p.a. *vs.* 2.4% p.a.). The other source of growth is soybean oil produced from the increased production and crushing of soybeans. Demand growth for vegetable oil is expected to slow down in the coming decade due to: a) reduced growth in per capita food use in developing countries, at 1.5% p.a. compared to 3.0% in the previous decade; and b) only slight increases in biodiesel production from vegetable oils at 1.5% p.a., due to the gradual fulfilment of biodiesel mandates.

Protein meal production and consumption is dominated by soybean meal. Compared to the past decade, consumption growth of protein meal slows down (2.2% p.a. *vs.* 3.9% p.a.), reflecting both slower growth in global livestock production and saturated levels of protein meal in Chinese feed rations. Chinese consumption of protein meal is projected to grow by 2.7% p.a. compared to 7.9% p.a. in the previous decade, still exceeding the growth rate of animal production however.

Growth in the world trade of soybeans is expected to slow down considerably in the next decade, compared to the previous decade. This development is directly linked to the projected slower growth in soybean crushing in the People's Republic of China (hereafter "China"). Because the growth in livestock production is expected to be concentrated in the main protein meal producing countries, domestic use of protein meal increases while trade will only expand slightly in the coming decade, resulting in a declining share of trade in world production.

Whereas soybean, other oilseeds and protein meal exports are dominated by the Americas, vegetable oil exports continue to be dominated by Indonesia and Malaysia (Figure 3.2). Vegetable oil, at 42%, is one of the agricultural commodities with the highest share of production that is traded. It is expected that this share remains stable throughout the projection.

Figure 3.2. **Exports of oilseeds and oilseed products by region**

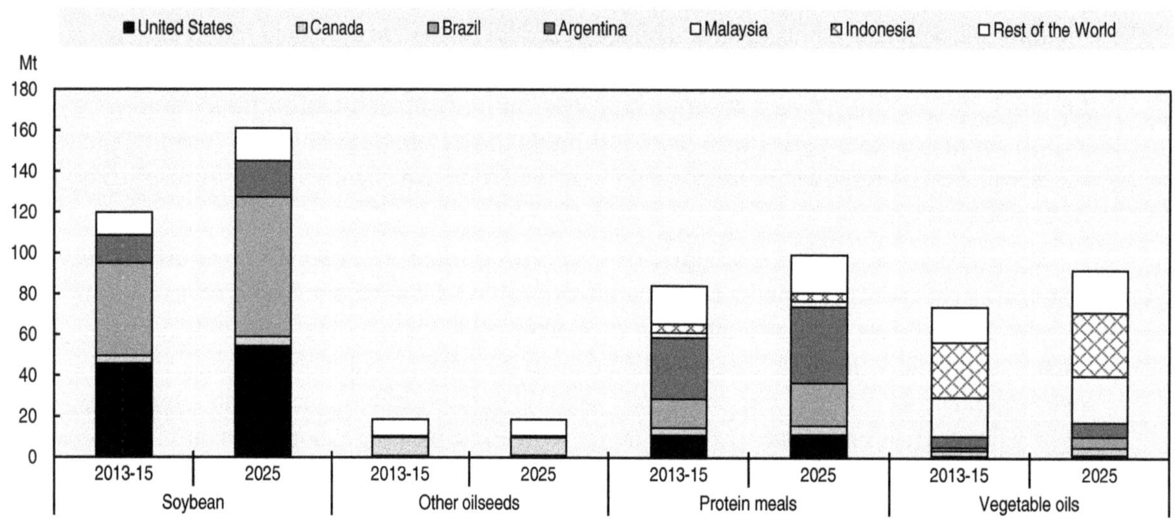

Source: OECD/FAO (2016), "OECD-FAO Agricultural Outlook", *OECD Agriculture statistics* (database), *http://dx.doi.org/10.1787/agr-data-en.*

StatLink *http://dx.doi.org/10.1787/888933381602*

In addition to the issues and uncertainties common to most commodities (e.g. macroeconomic environment, crude oil prices and weather conditions), the oilseed complex has its specific supply and demand sensitivities. The expected expansion of soybean and palm oil production depends on the availability of additional new land which might be constrained by new legislation aimed at protecting the environment. The low soybean stock-to-use level projected for the end of the outlook period is a source of uncertainty for the stability of prices if the sector is affected by adverse weather events. Biofuel policies in the United States, the European Union and Indonesia are also a source of major uncertainties in the vegetable oil sector because they have an impact on a considerable share of the demand in these countries.

The expanded oilseeds and oilseed products chapter is available at
http://dx.doi.org/10.1787/agr_outlook-2016-8-en

SUGAR

Market situation

International sugar prices fell by more than 30% in 2014. The prospect of a global sugar production deficit has led to a price increase at the start of the current season, but with stocks still at comfortable levels, the price increase is expected to average slightly above 2% in the 2015 marketing year (see glossary for a definition of marketing year).

Indeed, there has been a slowdown in output growth since 2013, and global sugar production is expected to fall by about 5 Mt in 2015. Given steady growth in global consumption, this should put an end to the surplus phase. Increases in sugar production are foreseen in Brazil (the leading producer and exporter), Australia, the Russian Federation and Thailand, but two main producers, India and the European Union, will see a decrease. After four years of replenishing global stocks, the stocks-to-use ratio should begin to decline at the start of the 2016-25 outlook period.

Projection highlights

The continuation of in place domestic policy measures as well as Brazil's sugarcane production prospects will continue to largely influence the sugar market over the medium-term. World sugar prices, when denominated in US dollars, are not expected to increase much as production prospects should be able to satisfy a growing world demand, notwithstanding WHO recommendations to reduce daily "free" sugar intake to less than 10% of total energy intake.

In terms of the macro-economic assumptions underpinning the *Outlook*, exchange rates are a key factor affecting the sugar market. Over the forecast period, the USD is assumed to strengthen against the majority of currencies, enhancing the competitiveness of major sugar exporters on the world market, especially Brazil. In contrast, a few deficit countries located mainly in Asia (China, Korea, Japan, Malaysia), will benefit from a firming of their exchange rates against the USD, making imports less expensive when denominated in local currencies.

World sugar prices, with the return to a deficit phase, are expected to increase only slightly for a couple of years as a consequence of high level of stocks and low oil prices. They are then foreseen to follow a moderate upward trend. The international raw sugar price (Intercontinental Exchange No. 11 contract nearby futures) is projected to reach USD 342/t (USD 15.5 cts/lb) in 2025, in nominal terms. Similarly, the indicator world white sugar price (Euronet, Liffe futures Contract No.407, London) is projected to reach USD 425/t (USD 19.2 cts/lb) in nominal terms in 2025. The white sugar premium (difference between white and raw sugar prices) should temporarily decline in 2017 with the decline in the EU raw sugar imports after quota abolition, before returning to a level close to USD 83/t at the end of the period.

The sugar sub-sectors in many developed and developing countries will continue to benefit from domestic policy support measures such as high import tariffs, tariff rate quotas, and minimum price support. These policies will continue to distort markets and contribute to the relatively elevated level of market volatility. However, new policies will

Figure 3.3. **Change in world nominal raw sugar prices when denominated in selected national currencies**

2025 vs. 2013-15

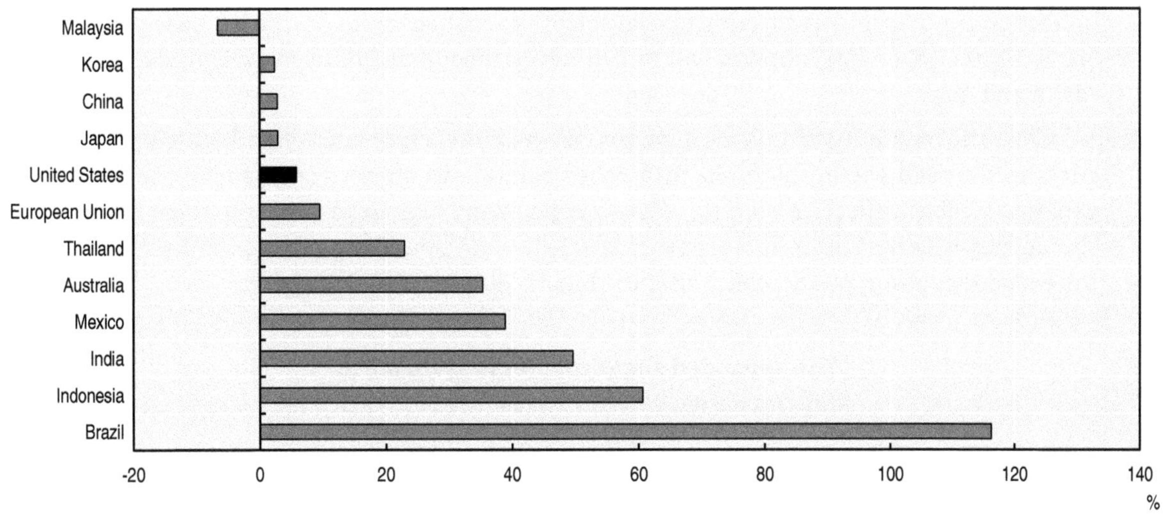

Source: OECD/FAO (2016), "OECD-FAO Agricultural Outlook", *OECD Agriculture statistics* (database), *http://dx.doi.org/10.1787/agr-data-en.*

StatLink http://dx.doi.org/10.1787/888933381613

liberalise the sugar market to some extent, such as the abolition of sugar quotas in 2017 in the European Union and the deregulation of sales of sugar in the open market in India.

Brazil's sugar sector has faced financial problems for several years, but will benefit from the weakness of the Brazilian real. Government policies continue to support ethanol production from sugarcane, but the share of sugarcane devoted to ethanol production should slightly decline over the outlook to 57%. This will displace sugar sales in domestic and export markets. Globally, a higher share of sugarcane production will be devoted to producing ethanol, rising from about 20.7% during the base period to 22.3% in 2025.

Global sugar production, despite an expected fall in the coming season in some producing countries, should rise over the course of the decade, sustained by demand growth and a reduction in stocks. Over the ten-year period, the growth in production is foreseen to average 2.1% per annum (p.a.), with production reaching 210 Mt by 2025, an increase of around 39 Mt over the base period (2013-15). Most of the additional production will originate in countries producing sugarcane rather than sugar beet, and the main driver of output growth is area expansion, notably in Brazil, even though yield improvements are foreseen for sugar crops and sugar processing in some other producing countries (India and Thailand).

The anticipated growth in world sugar demand for the next decade is steadier with an increase of 2% p.a. resulting in a decrease of the stock-to-use ratio from 45% in the base period to 39% in 2025. However, the growth in demand is mixed with nearly no growth in the matured developed countries and stronger prospects in developing countries, in particular Africa and Asia. In developing countries with high sugar calorie intake, no noticeable changes in consumer habits are foreseen, as sugar is an available, cheap source of energy, which is easy to transport and store.

In the face of growing global demand, sugar exports are likely to expand in countries that have modernised their sugar sub-sector in recent years (e.g. Australia, European Union, and Thailand). Brazil will remain the world's major producer and exporter, but lose market share at the start of the period, opting for more profitable ethanol production in the short-term. Favourable currency terms should encourage investment. Overall, Brazil's share of world sugar exports is expected to decline at the start of the projection period before recovering to a level close to that achieved during the base period (41%). On the other side, imports will remain diversified, mostly driven by demand from Africa and Asia.

Over the medium-term, the interaction between the sugar market and other sectors such as the feed sector, biofuels, and other caloric sweeteners (e.g. isoglucose) will generate feedback effects. Also, with existing policies and high fixed costs, the sugar sector should stay volatile. Furthermore, any external shock to one of the related markets, or to the exogenous assumptions, could alter the results discussed in this report.

The expanded sugar chapter is available at
http://dx.doi.org/10.1787/agr_outlook-2016-9-en

MEAT

Market situation

Weaker demand for meats by emerging economies and oil exporting countries throughout 2015 exerted significant downward pressure on meat prices. According to the FAO Meat Price Index, meat prices in 2015 fell to a level last seen in early 2010. This fall contrasts with an extended period of continued, though at times volatile, meat price increases that started back in 2002. Only once during this extended period – during the aftermath of the 2007-08 financial crisis – have meat prices fallen by such a magnitude.

World trade in 2015 stalled in volume terms. Meat exports from the Americas, the dominant supplier region, fell in 2015 reflecting weakening supply to the rest of the world. Lower imports from the Russian Federation, and a net trade loss in North America estimated at close to one million tonnes, substantially reduced supplies going to other parts of the world.

Projection highlights

The *Outlook* for the meat market remains strong. Feed grain prices are set to remain low for the projection period, giving stability to a sector that had been operating in an environment of particularly high and volatile feed costs over most of the past decade. This is particularly relevant for regions such as the Americas, Australia and Europe, where feed grains are being used more intensively in the production of meat.

Global meat production is projected to be 16% higher in 2025 than in the base period (2013-15). This compares with an increase of almost 20% in the previous decade. Developing countries are projected to account for the vast majority of the total increase, through a more intensive use of protein meal in feed rations. Poultry meat is the primary driver of the growth in total meat production in response to expanding global demand for this more affordable animal protein compared to red meats. Low production costs and lower product prices have contributed to making poultry the meat of choice both for producers and consumers in developing countries. In the bovine meat sector, several years of cow herd liquidation in major producing regions resulted in low beef production in 2015. However, production is expected to grow from 2016 onwards, with higher carcass weights more than offsetting the decline in cattle slaughter. Pigmeat production will also grow after 2016, driven by China, where herd size is expected to stabilise after years of substantial reductions (a drop of 25 million pigs between 2012 and 2015). Another factor contributing to China's output expansion in the coming years is further consolidation of the pork sector. Production is also expected to increase in the sheepmeat sector with an expected global growth of 2.1% p.a., a higher rate than the last decade, and led by China, Pakistan, Sudan and Australia.

Globally 10% of meat output will be traded in 2025, up from 9% in 2015, with most of the increase coming from poultry meat. Import demand will be weak during the first years of the outlook period, mainly due to lower imports due to the import ban of the Russian Federation and slower growth in China, but will strengthen in the second half of the projection period, due to import growth in the developing world. The most significant growth in import demand originates from Viet Nam, which captures a large share of additional imports for all meat types. Africa is another fast growing meat importing region albeit from a lower base. Although developed countries are still expected to account for slightly more than half of global meat exports by 2025, their share is steadily decreasing relative to the base period. On the other hand, Brazil's share of global exports is expected to increase to around 26%, contributing to nearly half of the expected increase in global meat exports over the projection period.

Nominal meat prices are expected to start at levels similar to those registered in 2010, and in most cases, trend marginally upwards. By 2025, prices for beef and pigmeat are projected to increase to around USD 4 497/t carcass weight equivalent (c.w.e.) and USD 1 580/t c.w.e. respectively, while world sheepmeat and poultry prices are expected to rise to around USD 3 857/t c.w.e. and USD 1 571/t product weight (p.w.) respectively. In real terms meat prices are expected to trend down from their recent high levels (Figure 3.4).

Figure 3.4. **World meat prices**

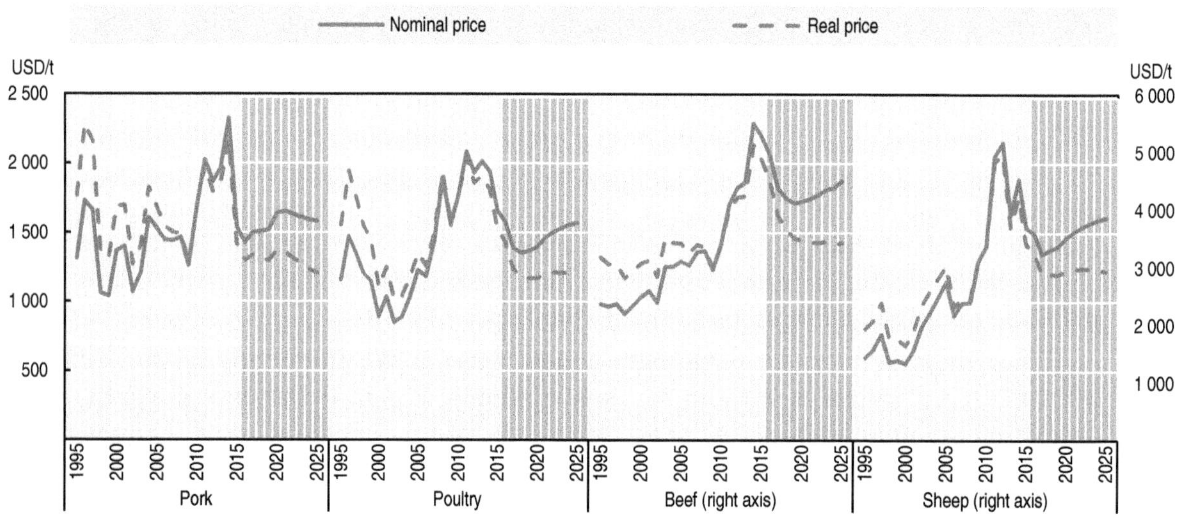

Note: US Choice steers, 1 100-1 300 lb dressed weight, Nebraska. New Zealand lamb schedule price dressed weight, all grade average. US Barrows and gilts, No. 1-3, 230-250 lb dressed weight, Iowa/South Minnesota. Brazil: Export unit value for chicken (FOB) product weight.
Source: OECD/FAO (2016), "OECD-FAO Agricultural Outlook", *OECD Agriculture statistics* (database), *http://dx.doi.org/10.1787/agr-data-en.*
StatLink *http://dx.doi.org/10.1787/888933381625*

Global annual meat consumption per capita is expected to reach 35.3 kg retail weight equivalent (r.w.e.) by 2025, an increase of 1.3 kg r.w.e. compared to the base period. This additional consumption will consist mainly of poultry. In absolute terms, total consumption growth in developed countries over the projection period is expected to remain small relative to developing regions, where rapid population growth and urbanisation remains the core drivers. This is particularly true in Sub-Saharan Africa, where the rate of total consumption growth over the outlook period is faster than any other region. The composition of growth is also somewhat unique, with the absolute growth in beef almost matching poultry.

Globally, animal disease outbreaks and trade policies remain among the main factors driving the evolution and dynamics in world meat markets. The implementation of various trade agreements, such as the proposed Trans-Pacific Partnership, over the outlook period could increase and diversify meat trade. An announcement in 2015 by International Agency for Research on Cancer of the World Health Organization (IARC) classified processed meat[2] as carcinogenic. This raised concerns among consumers worldwide and may impact the projected consumption of countries with high per capita meat consumption.

The expanded meat chapter is available at
http://dx.doi.org/10.1787/agr_outlook-2016-10-en

DAIRY AND DAIRY PRODUCTS

Market situation

International prices of all dairy products continued to decline from their 2013 peak, in particular for skim milk powder (SMP) and whole milk powder (WMP). A key factor was the decline in Chinese import demand, with demand for WMP dropping by 34% from 2014 levels. This decrease in Chinese demand for dairy products was coupled with continued production growth between 2014 and 2015, in key export markets, with total output of milk increasing in Australia (4%), the European Union (2%), New Zealand (5%) and the United States (1%).

The Russian Federation's ban on imports continues to restrict dairy trade. Russian cheese imports dropped by 62% between 2013 and 2015, which mainly affected exports from the European Union, the United States and Australia. Conversely, Belarus has greatly increased cheese exports to the Russian Federation, supplementing demand there. The ban is assumed to continue until the start of 2017; with imports of cheese expected to increase sharply as trading resumes, mostly supplied from the European Union and the United States, albeit at lower levels than prior to the ban.

Production in Oceania is facing challenges, low dairy prices have caused a reduction in the total dairy herd, which dropped by 2.7% in 2015. Furthermore drought and adverse weather conditions related to a very strong *El Niño* have restricted production in Oceania's pasture-based systems in 2016; this is expected to reduce production in New Zealand by 6.8% and to stall growth in Australia.

Previously good margins combined with the removal of the EU milk quota as of April 2015 has promoted growth in total milk production in the European Union. This growth, however, has been uneven across member states. For example, milk deliveries from the 2014 to 2015 marketing year (April-March) increased by 18.5% in Ireland, 3.7% in Germany, 2.9% in the United Kingdom, and 11.9% in the Netherlands. With increased milk production and limited growth in domestic consumption, EU exports for all major dairy commodities on aggregate are expected to increase by 58.5% between the 2013-15 base years and 2025.

Projection highlights

Per capita demand for dairy products in developing countries is expected to grow consistently over the medium-term, supported by rising incomes and lower dairy prices relative to their 2013 peak. As seen in previous years there is a continued shift in dietary patterns away from staples and towards animal products, due to changes in diets. Strong consumption growth is expected across several markets in the Middle East and Asia, including Saudi Arabia, Egypt, Iran and Indonesia, with the per capita consumption of dairy products in developing countries growing between 0.8% and 1.7% p.a., the lowest growth being for cheese and the highest for fresh dairy products. In addition, per capita consumption in the developed world is expected to grow between 0.5% for fresh dairy products and 1.1% p.a. for SMP.

World milk production is projected to increase by 177 Mt (23%) by 2025 compared to the base years (2013-15), corresponding to an average grow rate of 1.8% p.a. which is below the 2.0% p.a. witnessed in the last decade. The majority of this growth (73%) is anticipated to come from developing countries, in particular India and Pakistan. This expansion of production is largely in fresh dairy products, which will grow at 2.9% p.a. in developing countries, and predominantly supply domestic markets. At the world level, production of the main dairy products (butter, cheese, SMP and WMP) is increasing at similar pace to milk production, albeit more slowly than that of fresh dairy products.

As a result of these demand and supply factors, nominal prices of all dairy products are expected to increase over the medium-term, along with real prices of milk powders, which recover from their current lows. The real prices of butter and cheese are expected to decline slightly over the next decade, although this is from a comparatively higher base level than for milk powders.

The strengthening of both the US Dollar and the euro will put pressure on the growth of exports from the United States and the European Union, as they become less competitive on the global market. Conversely, exporters in Argentina, Australia and New Zealand are projected to become more competitive on the world market due to relatively weaker currencies.

Continued export growth is expected over the coming decade following the slump in 2014-15. Butter, cheese, SMP and whey all average strong growth of over 2%. Growth for exports of WMP is more modest at 1.8% p.a. With low dairy prices serving as a barrier to market entry for non-traditional exporters, export growth will continue to be satisfied by a small concentration of key exporters. The European Union will be the principle exporter of SMP and cheese, and New Zealand the lead exporter of butter and WMP, as shown in Figure 3.5.

Figure 3.5. **Exports of dairy products by region**

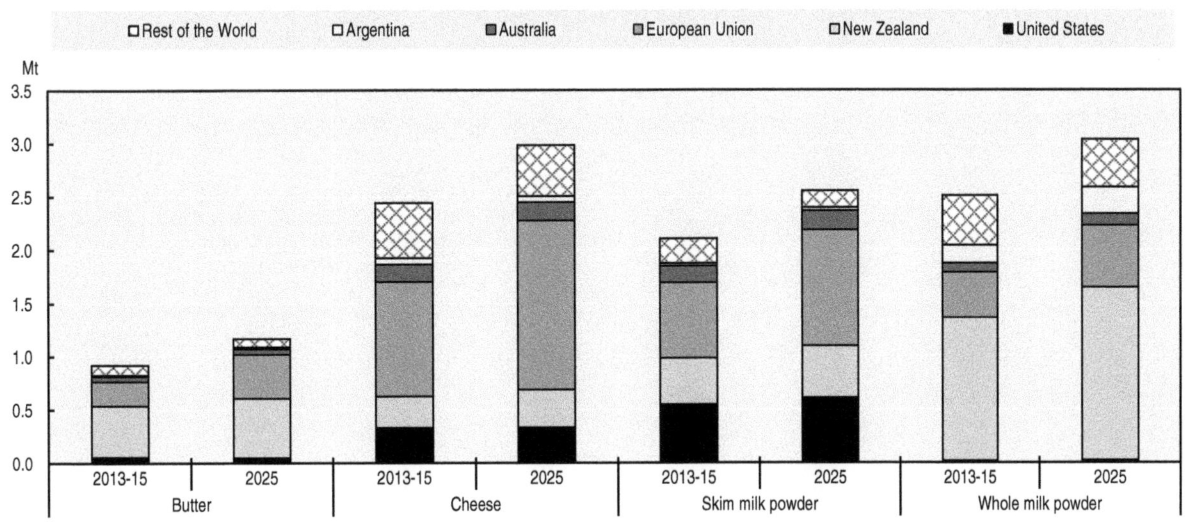

Source: OECD/FAO (2016), "OECD-FAO Agricultural Outlook", *OECD Agriculture statistics* (database), *http://dx.doi.org/10.1787/agr-data-en.*

StatLink *http://dx.doi.org/10.1787/888933381631*

The global dairy commodity market is prone to disturbances from weather variability, changes in policy, and the opening or closing of trade in key countries. Many of the developments in the dairy market will stem from import demand in the China and how quickly producers react to lower prices. The *Outlook* foresees a strengthening of demand from developing countries and assumes that while China does not resume importing WMP and butter at 2014 levels, instead servicing much of its demand internally, SMP and cheese imports will increase over the outlook period.

The expanded dairy and dairy products chapter is available at
http://dx.doi.org/10.1787/agr_outlook-2016-11-en

FISH AND SEAFOOD

Market situation

During 2015, the global fishery and aquaculture sector showed sustained growth in overall production and consumption. In 2014, aquaculture's contribution to total fish supplied for food overtook that of wild fish for the first time and this trend continued in 2015. In the same year, after a period of continuous expansion, trade of fish and fishery products declined in value terms. This slowdown was caused by economic contractions in key markets, exchange rate developments and lower fish prices. China, the leading producer, processor and exporter, and the third largest importer of fish and fishery products entered a period of serious uncertainty, even reducing its fish exports due to a slowdown in its processing sector. Seafood consumption in the Russian Federation suffered from the effects of its continuing trade embargo on fish from certain countries. Norway had record total export values, while in Thailand and other large shrimp supplying countries lower shrimp prices pushed total export values down significantly. Catches of anchoveta (mainly used to produce fishmeal and fish oil) were better than expected, relieving some short-term pressure on fishmeal and fish oil prices.

Prices of wild species increased more than those of farmed seafood in 2015, as measured by the FAO Fish Price Index (base 2002-04 = 100). Since reaching a peak in March 2014, with the index at 164, overall fish prices have shown a decreasing trend, with the index falling to 135 in July 2015 due to reduced consumer demand in key markets and an increased supply in certain fishery species. During the end of 2015 and early 2016, prices started to slightly recover.

Projection highlights

The outlook for the fish sector remains largely positive. In nominal terms, average fish prices are all expected to decline in the first part of the projection period before recovering in the last five years of the outlook period. In 2025, average producer prices are projected to be slightly higher than during the 2013-15 base period, as demand growth is expected to outpace supply. However, the average prices for traded products for human consumption, fishmeal and fish oil are projected to be slightly lower in 2025 relative to the base period. In real terms, however, all prices are expected to decrease over the next decade from the record highs attained in 2014.

World fish production is projected to grow at 1.5% p.a. during the outlook period, a slowdown relative to the 2.5% p.a. of the previous decade. Production is expected to reach 196 Mt, with an overall increase of 29 Mt, or 17%, between the base period and 2025. Most of the production growth for fish will take place in developing countries and in particular in Asia. As capture fisheries production is expected to increase by only 1%, by 2025, the majority of growth will come from aquaculture, which will surpass total capture fisheries in 2021 (Figure 3.6). Despite the increasing role of aquaculture in total fish supply, the capture sector is expected to remain dominant for a number of species and vital for domestic and international food security.

Aquaculture will continue to be one of the fastest growing food sectors despite its average annual growth rate slowing from 5.4% p.a. in the previous decade to 3.0% p.a. in the period 2016-25. This deceleration is due to higher costs, combined with competition for land, water and labour from alternative production systems. Much of the increase is expected in freshwater species.

Figure 3.6. **Aquaculture production and capture fisheries**

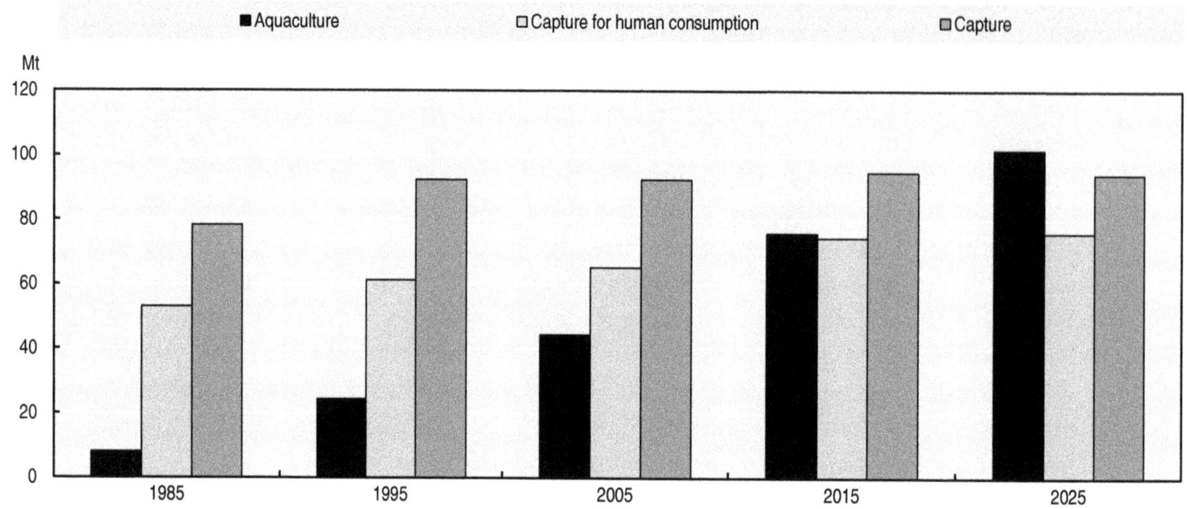

Note: "Capture for human consumption refers" to the Capture production excluding ornamental fish, fish destined to the production of fishmeal, fish oil and other non-food uses. All aquaculture production is assumed to be destined to human consumption.
Source: OECD/FAO (2016), "OECD-FAO Agricultural Outlook", *OECD Agriculture statistics* (database), *http://dx.doi.org/10.1787/agr-data-en*.
StatLink *http://dx.doi.org/10.1787/888933381641*

World production of fishmeal is expected to increase by 15% in 2025 relative to the average 2013-15 level to reach 5.1 Mt, while fish oil should grow by 17% to 1 Mt during the same period. Approximately 38% of fishmeal in 2025 will be obtained from fish by-products.

World fish consumption as food is projected to increase by 21% (or 31 Mt live weight (lw)) in 2025 compared to the base period, growing at 1.8% p.a. in the next decade compared to 3.1% p.a. in the previous one. In 2025, fish originating from aquaculture is expected to represent 57% of the fish consumed. Fish consumption will continue to expand more strongly in developing countries than developed countries, where there is an overall slowdown in consumption growth. Per capita fish consumption is expected to increase in all continents, while the fastest growth rates are projected for Oceania and Asia.

Fish and fishery products (fish for human consumption, fishmeal) will continue to be highly traded with about 36% of total fishery production (31% excluding intra-EU trade) expected to be exported in 2025. Trade of fish for human consumption is projected to increase by 18% (or 7 Mt lw) by 2025. However, its annual rate of growth is projected to decline from 2.3% p.a. during the last decade to 1.9% p.a. over the next decade reflecting the slowdown in production and demand. Developing countries will continue to be the main exporters of fish for human consumption, but their share in world exports will decrease from 67% in 2013-15 to 66% in 2025. During the same period, developed countries will reduce their share in world imports from 54% to 53%.

A number of uncertainties and challenges can affect projections for fish. The outlook for capture fisheries, fishmeal and fish oil depend on the natural productivity of fish stocks and ecosystems, which is uncertain, as well as on variable weather patterns. For aquaculture, relevant factors are the accessibility and availability of sites and water resources as well as to technology and finance; the sustainability, availability and cost of fish seeds (e.g. eggs, spawn, offspring, fry, larvae) and feeds; antibiotic use; assessment of environmental impacts (including pollution, fish diseases and escapees); and food safety and traceability issues. Furthermore, trade policies, trade agreements and market access remain important factors influencing the overall dynamics of world fish markets.

The expanded fish and seafood chapter is available at
http://dx.doi.org/10.1787/agr_outlook-2016-12-en

BIOFUELS

Market situation

Several political changes concerning biofuel markets were finalised in the course of 2015. In Brazil, the taxation system was amended to favour hydrous ethanol rather than gasohol[3] and the mandatory anhydrous ethanol blending ratio was increased from 25% to 27%. In the European Union, revisions to the Renewable Energy Directive (RED) and to the Fuel Quality Directive were adopted. A 7% cap was introduced on renewable energy coming from food and feed crops in the transport sector by 2020. After a long delay, the US Environmental Protection Agency's (EPA) final rulemaking for the years 2014-16 was issued in November 2015. The mandates specified are higher than those proposed earlier in the year, though still considerably lower than the initial levels proposed in 2007.

World ethanol[4] and biodiesel[5] prices continued to decrease in nominal terms in 2015 due to weak crude oil and biofuel feedstock prices. Demand for bioenergy in the transportation sector was mostly driven by blending mandates in major economies and by sustained fuel use around the world.

Projection highlights

International prices of ethanol and biodiesel are expected to recover in nominal terms over the outlook period, given developments in crude oil markets and the recovery of prices of biofuel feedstock (Figure 3.7). Global ethanol production is expected to expand modestly from 116 Bln L in 2015 to 128.4 Bln L by 2025. Half of this growth will originate from Brazil.

The expansion of global biodiesel production will be driven by policies in place in the United States, Argentina, Brazil and Indonesia, and to a lesser extent the fulfilment of the RED target in the European Union. It is expected to increase from 31 Bln L in 2015 to 41.4 Bln L by 2025. Advanced biofuels are not expected to take off over the projection period.

For the United States, this *Outlook* assumes that the 10% ethanol blend wall[6] will continue to limit growth in ethanol use, that biodiesel use will expand due to a stronger advanced mandate and that cellulosic ethanol will not be available on a large scale. The cellulosic mandate will be mostly met with renewable compressed natural gas and renewable liquefied natural gas. In the European Union, the proportion of total transport energy accounted for by biofuels, including double counting for sustainable biofuels is expected to reach 6.3% by 2020. The remainder of the 10% RED target will be met from other renewable energy sources such as electric cars.

In Brazil, it is assumed that prices will remain favourable to hydrous ethanol use rather than gasohol and thus a sustained demand for ethanol, mostly met by domestic production, will prevail over the outlook period. Indonesian biodiesel production will be used mainly to meet domestic mandate-driven demand. For India, the new policies aiming at compensating sugar mills for high sugar prices will encourage ethanol production from molasses.

Elsewhere in the world, development of the comparatively minor biofuels markets depends on a mix of effective policy support and price trends, leading to mixed prospects across countries.

Biofuel trade will remain limited. It is expected that ethanol exports will mostly originate from the United States where the blend wall limits further increases in domestic demand, and that biodiesel trade will be mostly directed from Argentina to the United States in order to meet the biodiesel and advanced mandates. Indonesian exports of biodiesel are expected to remain marginal given high tariffs imposed by importing countries.

The future evolution of energy markets as well as possible policy changes are key uncertainties attached to the *Outlook* for biofuel markets over the next decade. However, given recent policy decisions, uncertainties concerning the future of biofuel markets should ease somewhat, at least over the short-term.

Figure 3.7. **World biofuel prices**

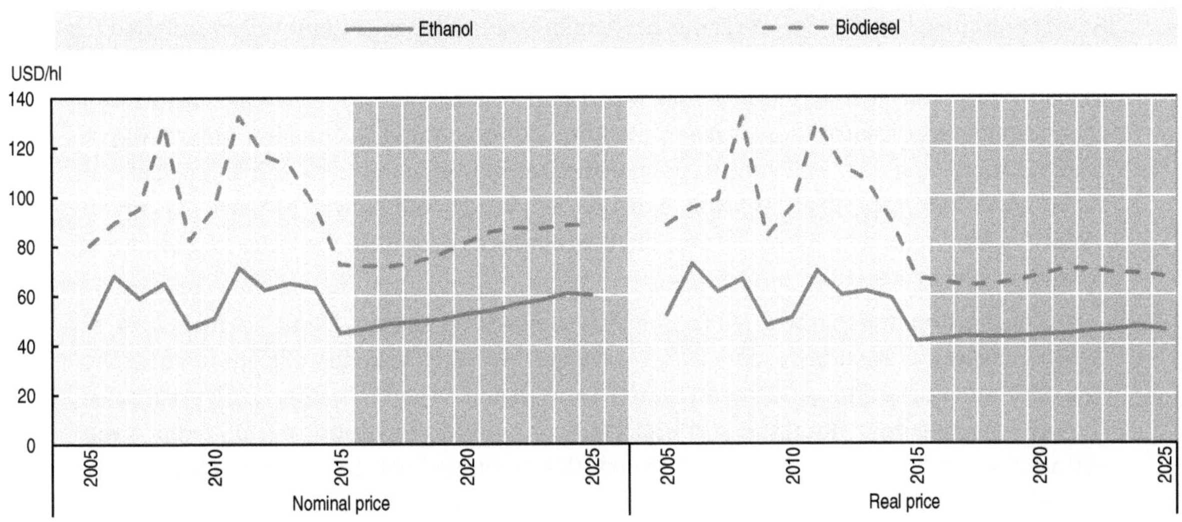

Note: Ethanol: wholesale price, US, Omaha; Biodiesel: Producer price, Germany, net of biodiesel tariff and energy tax.
Source: OECD/FAO (2016), "OECD-FAO Agricultural Outlook", *OECD Agriculture statistics* (database), *http://dx.doi.org/10.1787/agr-data-en.*
StatLink *http://dx.doi.org/10.1787/888933381654*

The expanded biofuels chapter is available at
http://dx.doi.org/10.1787/agr_outlook-2016-13-en

COTTON

Market situation

The world cotton market experienced dramatic developments in the first half of the 2015 marketing year (see glossary for a definition of marketing year) caused by an acute drop in production – about 9% – in major producing countries. Worldwide cotton production has not declined this much since 2008. This unexpected drop in production led to releases of stocks; however, total world stocks still remain at a very high level (20 Mt, 5% down from 2014).

Production fell in almost all major cotton producing countries led by Pakistan, the United States, and China, which experienced declines of 5%, 19% and 17%, respectively. Adverse weather, lower global world market demand and policy uncertainty all contributed to the sharp decline. The decreased synthetic fibre prices driven by substantially lower oil prices placed huge competitive pressures on world cotton markets. Nonetheless, cotton mill consumption is estimated to increase by 1% from 2014 to around 24.3 million tonnes (Mt) in the 2015 marketing year. Mill consumption estimates in China and India remained stable at 7.7 Mt and 5.3 Mt respectively, Pakistan experienced over 2% and Bangladesh over 4% growth while Viet Nam picked up 6% as Chinese direct investment in mills of the latter two countries continues to increase.

Global cotton imports declined for the third consecutive season, falling 2% from 2014, to 75 Mt. Increases in imports by Indonesia, Turkey and Viet Nam were insufficient to offset the 12% decline in China's import demand from 2014, as their new cotton support policy narrowed the price gap between domestic and imported cotton. With lower output, US exports are estimated to fall to 2.2 Mt, about 11% below the previous year. India's exports however increased slightly.

Projection highlights

Although the world cotton price is under pressure from substantial high stock levels and fierce competition from synthetic fibres, cotton prices are expected to be relatively stable in nominal terms after an anticipated further drop in 2016. During 2016-25, relative stability is expected as government support policies stabilise markets in major cotton producing countries. However, world cotton prices are expected to be lower than the average in 2013-15 in both real and nominal terms.

World production is expected to grow at slower pace than consumption during the first few years of the outlook period, reflecting the anticipated lower price level resulting from the large global stocks that accumulated between 2010 and 2014. The stock-to-use ratio is expected to be over 40% in 2025, which is at the high-end of historical levels but well below the historical high of 87% in 2014. World cotton area should be stable for the first five years but it is projected to grow from 2020 onwards. Yields rise around the world and global average yield grows slowly as production switches from relatively high yielding countries, notably China, to relatively low-yielding ones in South Asia.

World cotton use is expected to grow at 1.5% p.a. as a result of economic and population growth, reaching 28.3 Mt. Consumption in China is expected to fall to 6.9 Mt following the downward trend started in 2010, while India becomes the world's largest country for cotton

Figure 3.8. **Cotton consumption by region**

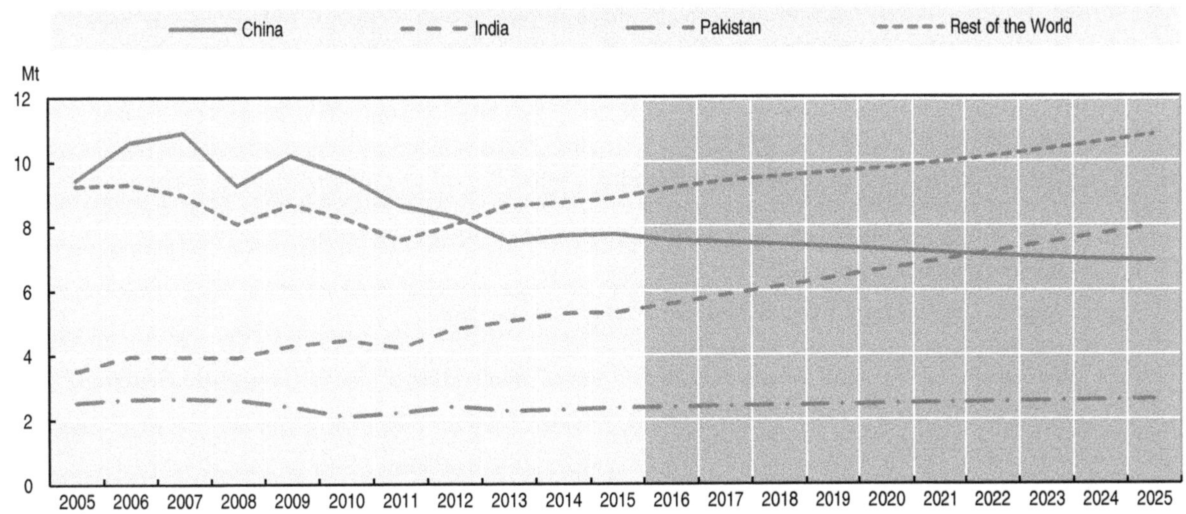

Source: OECD/FAO (2016), "OECD-FAO Agricultural Outlook", *OECD Agriculture statistics* (database), *http://dx.doi.org/10.1787/agr-data-en*.

StatLink *http://dx.doi.org/10.1787/888933381669*

mill consumption (8 Mt) in 2025. Higher cotton mill consumption by 2025 is also foreseen for Bangladesh, Pakistan, Turkey, Indonesia and Viet Nam.

It is expected that the growth in global cotton trade will be slower compared to previous years, especially 2011-13, when growth was driven by surging Chinese imports. To obtain value-added from mills, a shift to trading cotton yarn and fabrics rather than raw cotton has emerged over the past few years, which is expected to continue. Nonetheless, by 2025 global raw cotton trade will reach 8.7 Mt, nearly 7% higher than the average during 2013-15. The United States retains its position as the world's largest exporter, accounting for 28% of world trade. Exports from Brazil are expected to almost double from 0.7 Mt to 1.5 Mt, making it the world's second largest cotton exporter. With higher production, Australia is expected to increase cotton exports to 1.1 Mt, over 70% more than in the base period. Cotton producing countries in Sub-Saharan Africa, as a whole, will increase their exports to reach 1.4 Mt by 2025. On the import side, China is expected to import 1.6 Mt in 2025 and retains barely its position as the world's largest import market. Its dominant role in the world cotton market will be significantly challenged as other importing countries emerge. It is projected that by 2025, Bangladesh, Indonesia and Viet Nam will each import more than 1 Mt.

While continuing increases in farm labour costs and competition for resources with other agricultural crops place significant constraints on growth in global cotton production, higher productivity driven by technological progress, including greater adoption of bio-tech cotton, creates substantial potential for cotton production to expand in the next decade. While the medium-term prospects are for sustained growth, there may be potential short-term uncertainties in the current *Outlook* which may result in short-term volatilities in demand, supply and prices. A sudden slow-down in global economy, a sharp drop in global textiles and clothing trade, quality and price competition from synthetic fibres and changes in government policies are important factors that can affect the cotton market. The unprecedented high stock level is a key driver of the world cotton price.

The expanded cotton chapter is available at
http://dx.doi.org/10.1787/agr_outlook-2016-14-en

Notes

1. These absolute increases in the European Union and Canada are slightly misleading since the base periods included bumper crops in 2013 and 2014 in the European Union and an extreme bumper crop in Canada in 2013.
2. The term "processed meats" refers to meat that has been treated either to be preserved or flavoured, such as hams and sausages.
3. Gasohol is a mixture of gasoline and anhydrous ethanol used as transport fuel. In Brazil, most vehicles are flex-fuel vehicles able to run on any blend of gasoline and ethanol. At the pump, automobilists can choose between gasohol (currently E27) and hydrous ethanol (E100).
4. Wholesale price, US Omaha.
5. Producer price, Germany, net of biodiesel tariff and of energy tax.
6. The term blend wall refers to short run technical constraints that act as an impediment to increased ethanol use. It is assumed in this *Outlook* that US cars will not be able to consume gasohol with more than 10% of ethanol.

ANNEX

Commodity snapshot tables

Table 3.A1.1. **World cereal projections**

Marketing year

		Average 2013-15est	2016	2017	2018	2019	2020	2021	2022	2023	2024	2025
WHEAT												
World												
Production	Mt	720.3	721.7	733.2	739.8	747.3	753.9	760.1	767.7	775.6	783.6	791.3
Area	Mha	222.2	222.6	222.8	223.3	223.6	223.7	223.6	223.6	223.9	224.1	224.4
Yield	t/ha	3.24	3.24	3.29	3.31	3.34	3.37	3.40	3.43	3.46	3.50	3.53
Consumption	Mt	709.9	727.7	733.4	740.4	746.7	752.2	759.9	766.5	773.7	781.4	789.6
Feed use	Mt	133.4	140.6	141.1	143.1	146.1	148.3	150.5	152.7	155.1	157.9	161.3
Food use	Mt	487.7	497.2	503.5	508.6	513.4	518.1	523.3	528.2	532.8	537.9	543.8
Biofuel use	Mt	8.8	9.4	9.8	10.4	10.4	10.1	9.9	9.5	9.2	9.2	9.3
Other use	Mt	80.0	80.5	79.1	78.3	76.8	75.7	76.2	76.2	76.7	76.4	75.3
Exports	Mt	157.8	152.5	155.0	158.3	161.2	163.3	165.6	167.8	170.0	172.4	174.5
Closing stocks	Mt	196.2	196.6	196.4	195.8	196.4	198.1	198.3	199.5	201.4	203.5	205.3
Price[1]	USD/t	271.0	216.5	211.7	210.8	213.1	216.6	224.0	229.0	233.2	235.5	236.9
Developed countries												
Production	Mt	381.5	374.0	379.9	383.2	386.7	388.9	391.1	394.0	397.2	400.5	403.3
Consumption	Mt	266.9	270.6	271.5	273.4	273.0	272.9	274.2	275.0	276.5	278.1	279.5
Net trade	Mt	109.1	106.3	108.4	111.2	113.5	115.0	116.6	118.3	119.8	121.5	122.8
Closing stocks	Mt	69.5	73.8	73.8	72.4	72.6	73.6	73.9	74.6	75.6	76.5	77.5
Developing countries												
Production	Mt	338.8	347.7	353.3	356.5	360.6	365.0	369.1	373.7	378.4	383.1	388.0
Consumption	Mt	443.0	457.1	461.9	467.0	473.7	479.3	485.7	491.5	497.2	503.3	510.1
Net trade	Mt	-106.9	-106.3	-108.4	-111.2	-113.5	-115.0	-116.6	-118.3	-119.8	-121.5	-122.8
Closing stocks	Mt	126.7	122.9	122.6	123.3	123.8	124.5	124.4	124.8	125.8	127.0	127.7
OECD[2]												
Production	Mt	291.8	290.9	291.3	293.6	295.9	297.1	298.6	300.6	302.8	305.0	306.6
Consumption	Mt	219.2	222.9	223.9	225.4	225.0	224.9	225.8	226.4	227.8	228.9	229.8
Net trade	Mt	68.7	67.9	69.7	69.7	71.0	71.4	72.5	73.5	74.3	75.3	76.0
Closing stocks	Mt	51.9	57.0	54.7	53.2	53.2	54.0	54.3	54.9	55.7	56.5	57.3
MAIZE												
World												
Production	Mt	1 014.3	1 006.0	1 041.3	1 048.5	1 059.8	1 075.1	1 090.0	1 103.4	1 117.1	1 132.3	1 146.0
Area	Mha	181.1	178.9	182.7	182.3	182.4	183.0	183.5	183.8	184.1	184.6	184.8
Yield	t/ha	5.60	5.62	5.70	5.75	5.81	5.88	5.94	6.00	6.07	6.13	6.20
Consumption	Mt	986.7	1 024.9	1 038.6	1 060.1	1 062.8	1 073.7	1 087.2	1 102.6	1 114.7	1 130.6	1 143.4
Feed use	Mt	555.9	583.9	599.2	614.8	616.1	622.1	635.8	649.2	660.3	671.7	682.9
Food use	Mt	130.9	136.3	138.3	140.8	143.2	145.8	148.4	150.9	153.6	156.2	158.9
Biofuel use	Mt	148.0	156.9	161.4	160.9	159.6	159.4	157.1	156.3	155.2	156.6	154.4
Other use	Mt	102.0	101.4	93.3	96.0	95.4	97.2	96.7	96.1	95.4	95.4	95.8
Exports	Mt	130.2	128.6	127.9	129.1	130.2	131.0	133.1	134.7	137.4	139.8	141.5
Closing stocks	Mt	216.2	204.2	206.9	195.2	192.2	193.6	196.4	197.2	199.6	201.3	203.9
Price[3]	USD/t	180.6	170.2	163.0	163.0	165.9	170.0	175.9	181.2	182.2	183.9	186.7
Developed countries												
Production	Mt	498.0	488.6	506.4	507.9	511.3	516.1	521.3	525.5	530.2	536.0	541.5
Consumption	Mt	444.2	453.9	461.7	474.2	468.6	470.8	474.5	480.6	483.2	489.2	492.2
Net trade	Mt	45.0	37.6	38.9	40.4	41.3	41.3	42.7	44.3	46.3	47.6	49.2
Closing stocks	Mt	66.3	64.0	69.9	63.1	64.5	68.6	72.7	73.4	74.1	73.3	73.4
Developing countries												
Production	Mt	516.3	517.4	534.9	540.5	548.5	559.0	568.8	577.8	586.9	596.3	604.5
Consumption	Mt	542.5	571.1	576.9	585.9	594.2	603.0	612.7	622.0	631.6	641.4	651.2
Net trade	Mt	-37.9	-37.6	-38.9	-40.4	-41.3	-41.3	-42.7	-44.3	-46.3	-47.6	-49.2
Closing stocks	Mt	149.9	140.2	137.0	132.1	127.7	125.0	123.7	123.8	125.5	128.0	130.6
OECD[2]												
Production	Mt	466.2	463.1	474.9	475.6	478.4	482.6	487.2	490.8	494.8	499.9	504.8
Consumption	Mt	459.7	471.9	479.2	491.8	486.0	488.0	491.8	497.8	500.4	506.4	509.2
Net trade	Mt	-2.7	-5.3	-9.2	-9.2	-8.9	-9.7	-8.8	-7.7	-6.2	-5.5	-4.4
Closing stocks	Mt	65.6	63.7	68.6	61.6	63.0	67.2	71.4	72.0	72.7	71.7	71.7

StatLink http://dx.doi.org/10.1787/888933382148

Table 3.A1.1. **World cereal projections** *(cont.)*

Marketing year

		Average 2013-15est	2016	2017	2018	2019	2020	2021	2022	2023	2024	2025
OTHER COARSE GRAINS												
World												
Production	Mt	298.9	294.1	295.2	296.9	300.3	304.2	307.8	312.0	316.1	320.0	324.2
Area	Mha	158.2	157.4	156.9	156.7	157.1	157.7	158.0	158.6	159.2	159.6	160.2
Yield	t/ha	1.89	1.87	1.88	1.89	1.91	1.93	1.95	1.97	1.99	2.01	2.02
Consumption	Mt	290.5	291.2	294.0	296.8	299.9	303.0	306.6	310.1	313.9	317.7	321.6
Feed use	Mt	163.5	160.1	161.0	162.3	164.7	167.5	169.7	171.9	173.9	175.9	177.9
Food use	Mt	74.5	77.4	78.4	79.8	81.2	82.7	84.1	85.7	87.2	88.8	90.5
Biofuel use	Mt	3.9	4.2	4.2	4.3	4.3	4.2	3.8	3.6	3.4	3.4	3.4
Other use	Mt	48.6	49.6	50.4	50.3	49.7	48.7	49.0	49.0	49.3	49.5	49.9
Exports	Mt	45.5	40.2	40.4	40.7	41.4	42.6	43.6	45.1	46.6	48.1	49.8
Closing stocks	Mt	55.3	59.0	58.4	56.7	55.2	54.5	53.7	53.7	54.1	54.6	55.3
Price[4]	USD/t	212.4	156.4	153.9	157.1	161.4	166.0	175.3	181.0	184.5	188.4	190.0
Developed countries												
Production	Mt	185.8	177.5	177.3	176.6	177.4	178.6	179.7	181.0	182.3	183.5	184.9
Consumption	Mt	150.7	151.9	152.6	152.7	152.5	152.5	152.8	152.7	152.9	153.0	153.1
Net trade	Mt	31.3	24.8	24.9	25.2	26.2	27.1	27.8	28.7	29.6	30.5	31.7
Closing stocks	Mt	35.4	39.4	39.2	37.9	36.6	35.6	34.8	34.3	34.2	34.2	34.3
Developing countries												
Production	Mt	113.1	116.6	118.0	120.3	122.8	125.5	128.1	131.0	133.8	136.5	139.4
Consumption	Mt	139.8	139.3	141.5	144.0	147.3	150.6	153.9	157.4	161.0	164.7	168.6
Net trade	Mt	-26.7	-23.0	-23.1	-23.3	-24.4	-25.3	-25.9	-26.8	-27.7	-28.7	-29.8
Closing stocks	Mt	19.9	19.6	19.2	18.8	18.6	18.8	19.0	19.4	20.0	20.4	21.0
OECD[2]												
Production	Mt	155.4	148.4	147.9	147.3	148.0	149.1	150.0	151.2	152.4	153.4	154.6
Consumption	Mt	130.6	131.1	131.7	131.9	131.8	131.8	132.1	132.1	132.2	132.2	132.1
Net trade	Mt	21.5	17.0	16.6	16.6	17.4	18.2	18.7	19.6	20.4	21.3	22.4
Closing stocks	Mt	28.9	32.8	32.4	31.2	30.1	29.2	28.4	28.0	27.7	27.7	27.7
RICE												
World												
Production	Mt	493.4	503.6	510.5	517.7	524.7	531.0	537.0	543.4	549.7	556.3	562.6
Area	Mha	162.3	160.2	160.3	160.6	160.8	161.0	161.0	161.2	161.3	161.4	161.5
Yield	t/ha	3.04	3.14	3.18	3.22	3.26	3.30	3.34	3.37	3.41	3.45	3.48
Consumption	Mt	490.8	510.3	512.1	519.0	525.8	532.3	538.7	544.8	550.5	556.7	563.2
Feed use	Mt	20.9	22.6	22.5	22.6	22.7	22.9	23.1	23.2	23.4	23.5	23.5
Food use	Mt	397.9	408.8	413.9	419.6	424.9	430.2	435.1	439.8	444.1	448.7	453.6
Exports	Mt	44.3	42.7	43.7	44.8	46.2	47.1	48.0	48.7	49.6	50.6	51.4
Closing stocks	Mt	169.7	159.1	157.5	156.1	155.1	153.7	152.0	150.6	149.9	149.4	148.9
Price[5]	USD/t	454.7	399.6	407.6	409.2	405.8	407.0	409.7	413.4	415.5	415.4	416.3
Developed countries												
Production	Mt	17.9	18.0	18.2	18.4	18.5	18.6	18.7	18.8	18.9	19.0	19.1
Consumption	Mt	18.9	18.8	18.8	18.9	19.0	19.1	19.1	19.2	19.3	19.4	19.4
Net trade	Mt	-1.0	-0.5	-0.4	-0.4	-0.4	-0.5	-0.5	-0.5	-0.5	-0.5	-0.6
Closing stocks	Mt	5.3	4.8	4.6	4.5	4.4	4.4	4.4	4.5	4.6	4.8	5.1
Developing countries												
Production	Mt	475.4	485.6	492.2	499.3	506.2	512.4	518.3	524.6	530.9	537.2	543.5
Consumption	Mt	471.9	491.4	493.3	500.1	506.8	513.3	519.5	525.5	531.3	537.3	543.8
Net trade	Mt	1.8	0.5	0.4	0.4	0.4	0.5	0.5	0.5	0.5	0.5	0.6
Closing stocks	Mt	164.4	154.3	152.8	151.6	150.7	149.3	147.6	146.1	145.2	144.6	143.7
OECD[2]												
Production	Mt	21.6	21.4	21.6	21.7	21.7	21.8	21.9	21.9	22.0	22.0	22.1
Consumption	Mt	22.7	22.7	22.7	22.8	22.8	22.8	22.8	22.8	22.8	22.9	22.8
Net trade	Mt	-1.3	-0.9	-0.9	-0.9	-0.9	-0.9	-0.9	-0.9	-1.0	-1.0	-1.0
Closing stocks	Mt	6.7	6.5	6.2	6.1	5.9	5.9	5.9	5.9	6.0	6.1	6.4

Note: Marketing year: See Glossary of Terms for definitions.
Average 2013-15est: Data for 2015 are estimated.

1. No.2 hard red winter wheat, ordinary protein, United States FOB Gulf Ports (June/May), less EEP payments where applicable.
2. Excludes Iceland but includes all EU28 member countries.
3. No.2 yellow corn, United States FOB Gulf Ports (September/August).
4. Feed barley, Europe, FOB Rouen.
5. Milled 100%, grade b, nominal price quote, FOB Bangkok (January/December).

Source: OECD/FAO (2016), "OECD-FAO Agricultural Outlook", *OECD Agriculture statistics* (database). doi: dx.doi.org/10.1787/agr-outl-data-en

StatLink http://dx.doi.org/10.1787/888933382148

Table 3.A1.2. World oilseed projections

Marketing year

		Average 2013-15est	2016	2017	2018	2019	2020	2021	2022	2023	2024	2025
SOYBEAN												
World												
Production	Mt	298.7	318.7	324.5	334.7	342.9	351.3	358.9	367.9	377.6	384.4	393.9
Consumption	Mt	298.8	320.9	326.7	334.9	343.2	351.9	359.0	367.2	376.6	384.0	393.9
Crush	Mt	268.1	289.0	294.5	302.2	310.3	318.8	325.5	333.4	342.3	349.5	359.0
Closing stocks	Mt	34.6	36.2	33.9	33.7	33.5	32.9	32.7	33.4	34.4	34.8	34.8
Price[1]	USD/t	433.3	347.5	370.7	373.5	381.1	390.5	414.2	423.7	416.1	430.0	427.3
Developed countries												
Production	Mt	117.0	121.5	123.7	126.0	127.2	128.4	129.9	131.8	133.5	134.4	135.9
Consumption	Mt	83.3	87.2	87.7	88.7	89.2	90.1	90.7	91.0	92.1	92.3	93.4
Crush	Mt	75.3	79.3	79.7	80.5	81.4	82.3	82.9	83.2	84.2	84.6	85.7
Closing stocks	Mt	10.1	14.4	13.0	12.5	12.2	12.2	11.9	12.1	12.4	12.2	12.5
Developing countries												
Production	Mt	181.7	197.2	200.8	208.8	215.7	222.9	229.0	236.1	244.0	250.0	258.0
Consumption	Mt	215.5	233.7	239.0	246.2	254.0	261.8	268.3	276.2	284.5	291.7	300.5
Crush	Mt	192.8	209.7	214.8	221.6	229.0	236.5	242.6	250.2	258.1	264.9	273.3
Closing stocks	Mt	24.5	21.8	20.9	21.2	21.2	20.7	20.9	21.3	22.0	22.6	22.4
OECD[2]												
Production	Mt	110.3	114.0	115.5	117.5	118.4	119.3	120.5	122.0	123.4	124.0	125.1
Consumption	Mt	83.5	87.5	87.9	88.9	89.5	90.3	91.0	91.3	92.4	92.7	93.9
Crush	Mt	75.4	79.5	79.8	80.6	81.5	82.5	83.1	83.4	84.5	84.8	86.0
Closing stocks	Mt	9.9	14.3	12.9	12.3	12.1	12.0	11.7	12.0	12.3	12.1	12.3
OTHER OILSEEDS												
World												
Production	Mt	144.5	143.6	145.9	148.0	149.8	151.7	153.6	155.5	157.1	158.9	160.7
Consumption	Mt	143.9	143.4	145.1	147.4	149.5	151.5	153.5	155.7	157.3	159.1	160.7
Crush	Mt	119.6	118.8	120.5	122.6	124.7	126.6	128.6	130.7	132.4	134.2	135.8
Closing stocks	Mt	9.3	8.3	9.0	9.6	9.9	10.1	10.1	10.0	9.7	9.4	9.4
Price[3]	USD/t	444.7	400.5	409.0	391.6	391.0	401.9	412.5	407.2	411.8	422.3	433.1
Developed countries												
Production	Mt	83.8	82.7	84.3	85.4	86.5	87.5	88.6	89.7	90.6	91.8	92.8
Consumption	Mt	74.8	75.3	76.3	77.5	78.5	79.5	80.5	81.5	82.3	83.3	84.0
Crush	Mt	67.2	67.4	68.4	69.4	70.4	71.3	72.3	73.3	74.0	74.9	75.7
Closing stocks	Mt	7.0	6.9	7.6	8.1	8.4	8.6	8.6	8.4	8.1	7.8	7.8
Developing countries												
Production	Mt	60.7	60.9	61.6	62.6	63.3	64.2	64.9	65.8	66.5	67.1	67.9
Consumption	Mt	69.1	68.1	68.8	70.0	71.0	72.0	73.0	74.2	75.0	75.9	76.7
Crush	Mt	52.4	51.4	52.1	53.2	54.3	55.3	56.3	57.4	58.4	59.3	60.1
Closing stocks	Mt	2.3	1.4	1.4	1.5	1.5	1.5	1.5	1.6	1.6	1.6	1.6
OECD[2]												
Production	Mt	58.2	56.0	57.0	57.4	57.9	58.3	58.6	59.1	59.3	59.8	60.2
Consumption	Mt	54.5	53.9	54.4	54.9	55.3	55.7	56.0	56.4	56.6	56.9	57.1
Crush	Mt	48.6	47.7	48.2	48.6	49.0	49.3	49.7	50.0	50.2	50.5	50.7
Closing stocks	Mt	5.8	5.5	6.2	6.6	6.8	7.0	7.0	6.7	6.4	6.1	6.0
PROTEIN MEALS												
World												
Production	Mt	301.1	317.0	322.7	330.6	338.6	346.8	353.6	361.5	369.9	377.0	385.7
Consumption	Mt	298.0	316.9	322.9	330.6	338.5	346.5	353.6	361.2	369.6	376.8	385.6
Closing stocks	Mt	16.0	16.1	15.9	16.0	16.1	16.4	16.4	16.7	16.9	17.1	17.3
Price[4]	USD/t	391.0	289.7	296.9	302.8	312.3	318.0	340.8	350.0	352.1	360.8	368.5
Developed countries												
Production	Mt	99.4	102.0	103.0	104.5	105.7	106.9	107.9	108.6	109.7	110.4	111.6
Consumption	Mt	114.6	118.9	119.4	120.9	122.5	123.7	124.3	125.2	126.4	127.0	128.3
Closing stocks	Mt	2.2	2.0	2.0	2.1	2.1	2.2	2.2	2.2	2.3	2.3	2.4
Developing countries												
Production	Mt	201.8	215.0	219.7	226.1	232.9	239.9	245.8	252.9	260.2	266.6	274.2
Consumption	Mt	183.4	198.0	203.4	209.7	216.1	222.8	229.3	236.0	243.2	249.9	257.2
Closing stocks	Mt	13.8	14.1	13.9	13.9	14.0	14.2	14.2	14.4	14.7	14.8	14.9
OECD[2]												
Production	Mt	92.5	94.8	95.5	96.8	97.8	98.7	99.5	100.0	101.0	101.4	102.4
Consumption	Mt	119.2	123.2	123.8	125.3	126.7	127.9	128.5	129.5	130.9	131.6	133.2
Closing stocks	Mt	2.1	2.0	1.9	1.9	1.9	2.0	2.0	2.0	2.0	2.0	2.0

StatLink http://dx.doi.org/10.1787/888933382156

Table 3.A1.2. **World oilseed projections** *(cont.)*

Marketing year

		Average 2013-15est	2016	2017	2018	2019	2020	2021	2022	2023	2024	2025
VEGETABLE OILS												
World												
Production	Mt	174.1	180.5	184.2	188.5	192.7	197.1	201.2	205.6	210.1	214.3	218.9
of which palm oil	Mt	61.1	63.3	65.1	66.8	68.4	70.0	71.7	73.4	75.1	76.8	78.6
Consumption	Mt	173.4	181.0	184.1	187.8	192.1	196.7	200.7	204.9	209.4	213.8	218.3
Food	Mt	141.6	147.0	149.4	152.8	156.1	159.4	162.6	166.3	170.2	173.9	178.0
Biofuel	Mt	22.4	23.3	23.5	23.6	24.2	25.1	25.4	25.7	25.8	26.2	26.2
Exports	Mt	74.1	76.7	78.0	79.5	81.3	83.0	84.6	86.4	88.3	90.3	92.1
Closing stocks	Mt	23.7	22.5	22.7	23.3	23.9	24.3	24.8	25.5	26.2	26.7	27.2
Price[5]	USD/t	782.2	736.5	759.8	761.9	777.2	806.0	826.6	826.5	821.1	830.3	834.3
Developed countries												
Production	Mt	44.0	44.8	45.2	45.8	46.4	47.0	47.5	48.0	48.6	49.1	49.7
Consumption	Mt	49.1	49.1	49.1	49.2	49.5	50.1	50.2	50.3	50.3	50.4	50.2
Closing stocks	Mt	3.9	3.9	3.9	3.9	3.9	3.9	4.0	4.0	4.1	4.0	4.1
Developing countries												
Production	Mt	130.1	135.8	139.0	142.7	146.4	150.1	153.7	157.6	161.5	165.2	169.2
Consumption	Mt	124.3	131.9	135.0	138.7	142.6	146.6	150.5	154.6	159.1	163.5	168.1
Closing stocks	Mt	19.8	18.6	18.8	19.4	20.0	20.3	20.8	21.4	22.1	22.6	23.1
OECD[2]												
Production	Mt	36.5	36.9	37.1	37.5	37.8	38.1	38.4	38.6	39.0	39.2	39.5
Consumption	Mt	48.8	49.2	49.2	49.3	49.8	50.4	50.6	50.7	50.8	51.0	50.8
Closing stocks	Mt	3.3	3.3	3.3	3.3	3.3	3.3	3.4	3.4	3.4	3.4	3.5

Note: Average 2013-15est: Data for 2015 are estimated.

1. Soybean, U.S., CIF Rotterdam.
2. Excludes Iceland but includes all EU28 member countries.
3. Rapeseed, Europe, CIF Hamburg.
4. Weighted average protein meal, European port.
5. Weighted average price of oilseed oils and palm oil, European port.

Source: OECD/FAO (2016), "OECD-FAO Agricultural Outlook", *OECD Agriculture statistics* (database). doi: dx.doi.org/10.1787/agr-outl-data-en

StatLink http://dx.doi.org/10.1787/888933382156

Table 3.A1.3. **World sugar projections**

Marketing year

		Average 2013-15est	2016	2017	2018	2019	2020	2021	2022	2023	2024	2025
WORLD												
SUGARBEET												
Production	Mt	257.0	268.2	271.9	274.2	276.8	278.9	279.7	280.0	280.4	282.1	284.1
Area	Mha	4.4	4.5	4.5	4.5	4.6	4.6	4.5	4.5	4.5	4.5	4.5
Yield	t/ha	58.63	58.99	59.79	60.31	60.71	61.16	61.51	61.88	62.28	62.69	63.10
Biofuel use	Mt	12.8	12.7	10.0	10.5	10.4	10.4	10.4	9.5	9.5	9.3	9.3
SUGARCANE												
Production	Mt	1 811.6	1 850.3	1 875.1	1 908.5	1 943.3	1 976.0	2 009.2	2 046.4	2 083.4	2 118.0	2 151.9
Area	Mha	26.9	27.2	27.3	27.7	28.0	28.3	28.7	29.1	29.4	29.7	30.0
Yield	t/ha	67.37	68.08	68.57	68.98	69.41	69.74	70.03	70.42	70.86	71.30	71.73
Biofuel use	Mt	374.3	409.5	414.0	420.0	430.3	440.1	448.4	457.1	462.9	471.4	479.0
SUGAR												
Production	Mt tq	170.8	173.7	177.8	181.9	187.2	191.4	194.7	198.8	202.6	206.2	210.0
Consumption	Mt tq	166.8	172.9	175.3	178.2	181.8	185.7	189.6	193.5	197.3	201.0	204.7
Closing stocks	Mt tq	74.5	68.4	67.4	67.5	69.3	71.4	72.9	74.6	76.3	77.9	79.6
Price, raw sugar[1]	USD/t	323.0	325.3	330.4	337.6	335.8	330.9	334.0	333.1	333.5	337.4	341.9
Price, white sugar[2]	USD/t	408.0	398.6	388.3	402.5	405.8	407.0	406.9	408.8	413.7	419.0	424.5
Price, HFCS[3]	USD/t	539.7	472.1	450.8	457.6	470.1	477.2	476.3	487.7	498.6	500.6	508.3
DEVELOPED COUNTRIES												
SUGARBEET												
Production	Mt	202.2	210.8	213.5	214.8	216.1	216.9	216.6	215.8	215.1	215.8	216.7
SUGARCANE												
Production	Mt	78.1	82.9	85.0	86.9	89.2	90.4	90.8	91.0	91.5	92.3	93.1
SUGAR												
Production	Mt tq	39.1	40.2	41.5	41.9	42.6	42.9	43.0	43.2	43.4	43.7	43.9
Consumption	Mt tq	47.6	47.4	47.0	46.7	47.0	47.3	47.5	47.8	48.1	48.3	48.6
Closing stocks	Mt tq	14.1	12.4	12.7	12.9	13.4	13.9	14.2	14.4	14.5	14.5	14.5
HFCS												
Production	Mt	9.5	9.4	10.1	10.3	10.4	10.5	10.7	10.8	10.9	11.0	11.1
Consumption	Mt	8.2	8.1	8.7	8.9	9.0	9.0	9.2	9.2	9.3	9.3	9.3
DEVELOPING COUNTRIES												
SUGARBEET												
Production	Mt	54.8	57.4	58.4	59.4	60.7	62.0	63.2	64.2	65.3	66.3	67.4
SUGARCANE												
Production	Mt	1 733.5	1 767.4	1 790.2	1 821.7	1 854.0	1 885.6	1 918.4	1 955.3	1 992.0	2 025.7	2 058.8
SUGAR												
Production	Mt tq	131.8	133.5	136.4	140.0	144.6	148.5	151.7	155.6	159.3	162.6	166.1
Consumption	Mt tq	119.2	125.5	128.3	131.4	134.8	138.4	142.0	145.7	149.2	152.6	156.2
Closing stocks	Mt tq	60.3	56.0	54.7	54.6	55.9	57.5	58.7	60.2	61.8	63.4	65.1
HFCS												
Production	Mt	3.3	3.4	3.4	3.5	3.5	3.6	3.6	3.6	3.7	3.7	3.8
Consumption	Mt	4.1	4.3	4.4	4.5	4.6	4.7	4.7	4.8	4.9	5.0	5.1
OECD[4]												
SUGARBEET												
Production	Mt	161.3	166.8	169.9	171.1	172.5	173.1	172.7	172.1	171.6	172.0	172.7
SUGARCANE												
Production	Mt	120.3	125.0	126.2	128.0	130.8	132.3	132.5	132.3	132.4	133.0	134.1
SUGAR												
Production	Mt tq	37.9	38.9	40.2	40.6	41.2	41.5	41.6	41.7	41.8	42.0	42.3
Consumption	Mt tq	43.8	43.9	43.4	43.2	43.5	43.8	44.0	44.2	44.5	44.7	44.9
Closing stocks	Mt tq	12.4	11.2	11.3	11.2	11.4	11.7	11.7	11.7	11.6	11.6	11.6
HFCS												
Production	Mt	10.6	10.5	11.1	11.3	11.5	11.6	11.8	11.9	12.0	12.1	12.2
Consumption	Mt	10.3	10.2	10.8	11.1	11.2	11.3	11.5	11.6	11.7	11.9	11.9

Note: Marketing year: See Glossary of Terms for definitions.
Average 2013-15est: Data for 2015 are estimated.
tq : tel quel.
HFCS: High fructose corn syrup.

1. Raw sugar world price, ICE contract No11 nearby, October/September.
2. Refined sugar price, White Sugar Futures Contract No. 407, Euronext market, Liffe, London, Europe, October/September.
3. United States wholesale list price HFCS-55, October/September.
4. Excludes Iceland but includes all EU28 member countries.

Source: OECD/FAO (2016), "OECD-FAO Agricultural Outlook", *OECD Agriculture statistics* (database). doi: dx.doi.org/10.1787/agr-outl-data-en

StatLink *http://dx.doi.org/10.1787/888933382161*

Table 3.A1.4. **World meat projections**

Calendar year

		Average 2013-15est	2016	2017	2018	2019	2020	2021	2022	2023	2024	2025
WORLD												
BEEF AND VEAL												
Production	kt cwe	67 962	69 106	70 029	70 891	72 239	73 141	74 185	75 164	75 974	76 894	77 766
Consumption	kt cwe	67 340	68 832	69 718	70 598	71 934	72 836	73 874	74 851	75 666	76 586	77 461
PIGMEAT												
Production	kt cwe	117 005	118 402	119 920	121 533	123 045	124 468	125 838	127 115	128 528	129 833	131 001
Consumption	kt cwe	116 674	118 398	119 742	121 337	122 833	124 261	125 640	126 914	128 325	129 628	130 797
POULTRY MEAT												
Production	kt rtc	110 280	115 192	117 630	119 321	120 886	122 758	124 393	126 059	127 889	129 568	131 255
Consumption	kt rtc	109 811	115 247	117 608	119 321	120 895	122 766	124 392	126 058	127 888	129 565	131 255
SHEEP MEAT												
Production	kt cwe	14 137	14 473	14 778	15 076	15 426	15 800	16 229	16 571	16 844	17 113	17 438
Consumption	kt cwe	14 042	14 492	14 805	15 107	15 453	15 818	16 229	16 562	16 844	17 105	17 430
TOTAL MEAT												
Per capita consumption[1]	kg rwt	34.0	34.3	34.4	34.6	34.7	34.8	34.9	35.0	35.1	35.2	35.3
DEVELOPED COUNTRIES												
BEEF AND VEAL												
Production	kt cwe	29 198	29 277	29 441	29 782	30 214	30 380	30 580	30 786	30 916	31 080	31 246
Consumption	kt cwe	28 656	28 708	28 805	29 142	29 510	29 687	29 882	30 071	30 187	30 358	30 501
PIGMEAT												
Production	kt cwe	42 524	44 141	44 396	44 730	45 090	45 201	45 306	45 430	45 625	45 865	46 085
Consumption	kt cwe	39 608	40 864	41 156	41 381	41 666	41 741	41 823	41 871	41 960	42 080	42 178
POULTRY MEAT												
Production	kt rtc	45 679	47 975	48 781	49 341	49 821	50 367	50 825	51 235	51 694	52 128	52 575
Consumption	kt rtc	43 149	45 555	46 171	46 578	46 903	47 446	47 816	48 203	48 652	49 000	49 382
SHEEP MEAT												
Production	kt cwe	3 369	3 283	3 306	3 325	3 373	3 423	3 477	3 516	3 540	3 576	3 607
Consumption	kt cwe	2 631	2 650	2 664	2 667	2 682	2 697	2 716	2 735	2 754	2 769	2 783
TOTAL MEAT												
Per capita consumption[1]	kg rwt	65.0	66.8	67.2	67.5	67.8	68.1	68.3	68.5	68.7	68.9	69.2
DEVELOPING COUNTRIES												
BEEF AND VEAL												
Production	kt cwe	38 764	39 830	40 588	41 109	42 025	42 760	43 605	44 377	45 059	45 815	46 520
Consumption	kt cwe	38 685	40 124	40 913	41 456	42 424	43 149	43 991	44 780	45 479	46 228	46 959
PIGMEAT												
Production	kt cwe	74 481	74 261	75 524	76 803	77 954	79 267	80 532	81 685	82 904	83 969	84 916
Consumption	kt cwe	77 066	77 534	78 586	79 956	81 166	82 520	83 817	85 043	86 365	87 549	88 618
POULTRY MEAT												
Production	kt rtc	64 601	67 218	68 849	69 980	71 065	72 392	73 568	74 824	76 195	77 439	78 680
Consumption	kt rtc	66 661	69 692	71 438	72 743	73 991	75 320	76 577	77 855	79 236	80 564	81 873
SHEEP MEAT												
Production	kt cwe	10 768	11 190	11 472	11 751	12 053	12 377	12 752	13 055	13 304	13 537	13 831
Consumption	kt cwe	11 411	11 841	12 141	12 440	12 771	13 121	13 513	13 827	14 089	14 337	14 648
TOTAL MEAT												
Per capita consumption[1]	kg rwt	26.6	26.6	26.8	27.0	27.1	27.3	27.4	27.6	27.7	27.8	28.0
OECD[2]												
BEEF AND VEAL												
Production	kt cwe	27 289	27 419	27 621	28 008	28 404	28 523	28 695	28 878	28 966	29 093	29 208
Consumption	kt cwe	26 248	26 571	26 612	26 933	27 272	27 398	27 566	27 733	27 798	27 939	28 031
PIGMEAT												
Production	kt cwe	40 665	41 964	42 062	42 393	42 750	42 865	42 982	43 082	43 250	43 450	43 634
Consumption	kt cwe	37 913	39 399	39 392	39 637	39 922	40 011	40 129	40 187	40 292	40 421	40 526
POULTRY MEAT												
Production	kt rtc	44 110	46 238	47 080	47 661	48 143	48 712	49 171	49 579	50 043	50 510	51 000
Consumption	kt rtc	41 113	43 688	44 284	44 669	44 971	45 493	45 829	46 169	46 571	46 914	47 314
SHEEP MEAT												
Production	kt cwe	2 708	2 623	2 633	2 650	2 688	2 730	2 773	2 801	2 816	2 845	2 870
Consumption	kt cwe	1 990	2 003	2 002	2 002	2 007	2 015	2 026	2 036	2 047	2 054	2 062
TOTAL MEAT												
Per capita consumption[1]	kg rwt	65.7	67.8	67.9	68.2	68.4	68.6	68.7	68.7	68.8	69.0	69.1

Note: Calendar Year: Year ending 30 September for New Zealand.
Average 2013-15est: Data for 2015 are estimated.

1. Per capita consumption expressed in retail weight. Carcass weight to retail weight conversion factors of 0.7 for beef and veal, 0.78 for pigmeat and 0.88 for both sheep meat and poultry meat.
2. Excludes Iceland but includes all EU28 member countries.

Source: OECD/FAO (2016), "OECD-FAO Agricultural Outlook", *OECD Agriculture statistics* (database). doi: dx.doi.org/10.1787/agr-outl-data-en

StatLink http://dx.doi.org/10.1787/888933382176

Table 3.A1.5. **World dairy projections: Butter and cheese**

Calendar year

		Average 2013-15est	2016	2017	2018	2019	2020	2021	2022	2023	2024	2025
BUTTER												
World												
Production	kt pw	10 375	10 726	10 950	11 149	11 328	11 527	11 717	11 917	12 116	12 319	12 522
Consumption	kt pw	10 291	10 641	10 864	11 076	11 264	11 461	11 652	11 851	12 052	12 258	12 462
Stock changes	kt pw	-2	-4	14	4	-4	-5	-4	-1	-2	-2	-1
Price[1]	USD/t	3 650	2 848	2 969	2 968	3 051	3 188	3 310	3 356	3 405	3 482	3 520
Developed countries												
Production	kt pw	4 592	4 800	4 893	4 977	5 033	5 103	5 153	5 220	5 287	5 358	5 433
Consumption	kt pw	4 038	4 204	4 280	4 343	4 383	4 429	4 469	4 515	4 560	4 608	4 654
Developing countries												
Production	kt pw	5 782	5 925	6 057	6 172	6 295	6 424	6 564	6 697	6 829	6 961	7 089
Consumption	kt pw	6 253	6 436	6 583	6 733	6 882	7 033	7 183	7 336	7 493	7 650	7 807
OECD[2]												
Production	kt pw	4 245	4 450	4 533	4 604	4 651	4 723	4 778	4 853	4 921	4 987	5 057
Consumption	kt pw	3 634	3 841	3 893	3 955	3 995	4 042	4 083	4 134	4 184	4 233	4 281
Stock changes	kt pw	-2	-4	14	4	-4	-5	-4	-1	-2	-2	-1
CHEESE												
World												
Production	kt pw	22 224	22 937	23 366	23 747	24 031	24 382	24 722	25 039	25 350	25 696	26 028
Consumption	kt pw	21 918	22 692	23 145	23 471	23 773	24 116	24 454	24 772	25 084	25 431	25 765
Stock changes	kt pw	25	-22	-46	10	-8	0	3	1	1	1	-1
Price[3]	USD/t	4 064	3 154	3 275	3 328	3 392	3 507	3 558	3 627	3 698	3 782	3 834
Developed countries												
Production	kt pw	17 709	18 350	18 652	18 974	19 196	19 469	19 753	19 998	20 242	20 519	20 779
Consumption	kt pw	16 830	17 422	17 755	17 958	18 160	18 382	18 607	18 818	19 026	19 265	19 473
Developing countries												
Production	kt pw	4 515	4 586	4 715	4 774	4 835	4 913	4 969	5 041	5 108	5 177	5 249
Consumption	kt pw	5 088	5 270	5 390	5 512	5 613	5 734	5 846	5 954	6 057	6 166	6 292
OECD[2]												
Production	kt pw	17 079	17 602	17 924	18 242	18 457	18 718	19 003	19 249	19 489	19 755	20 004
Consumption	kt pw	16 250	16 930	17 194	17 397	17 596	17 813	18 030	18 233	18 434	18 671	18 876
Stock changes	kt pw	25	-22	-46	10	-8	0	3	1	1	1	-1

Note: Calendar year: Year ending 30 June for Australia and 31 May for New Zealand in OECD aggregate.
Average 2013-15est: Data for 2015 are estimated.

1. FOB export price, butter, 82% butterfat, Oceania.
2. Excludes Iceland but includes all EU28 member countries.
3. FOB export price, cheddar cheese, 39% moisture, Oceania.

Source: OECD/FAO (2016), "OECD-FAO Agricultural Outlook", *OECD Agriculture statistics* (database). doi: dx.doi.org/10.1787/agr-outl-data-en

StatLink http://dx.doi.org/10.1787/888933382182

Table 3.A1.6. **World dairy projections: Powders and casein**

Calendar year

		Average 2013-15est	2016	2017	2018	2019	2020	2021	2022	2023	2024	2025
SKIM MILK POWDER												
World												
Production	kt pw	4 181	4 449	4 351	4 421	4 505	4 626	4 741	4 830	4 952	5 061	5 169
Consumption	kt pw	4 013	4 341	4 466	4 503	4 544	4 652	4 751	4 850	4 953	5 067	5 177
Stock changes	kt pw	13	1	-17	-4	-2	4	12	4	5	2	2
Price[1]	USD/t	3 439	2 227	2 213	2 352	2 432	2 612	2 674	2 800	2 928	3 033	3 116
Developed countries												
Production	kt pw	3 625	3 858	3 759	3 817	3 886	3 985	4 087	4 158	4 262	4 353	4 447
Consumption	kt pw	1 871	1 999	2 036	2 009	1 984	2 028	2 054	2 084	2 118	2 158	2 196
Developing countries												
Production	kt pw	557	591	592	604	619	640	654	672	690	707	722
Consumption	kt pw	2 142	2 342	2 429	2 494	2 560	2 624	2 697	2 766	2 836	2 909	2 981
OECD[2]												
Production	kt pw	3 418	3 626	3 536	3 589	3 658	3 750	3 854	3 924	4 026	4 116	4 211
Consumption	kt pw	1 937	2 085	2 132	2 112	2 093	2 144	2 177	2 214	2 255	2 302	2 347
Stock changes	kt pw	13	1	-17	-4	-2	4	12	4	5	2	2
WHOLE MILK POWDER												
World												
Production	kt pw	4 011	4 121	4 269	4 375	4 465	4 549	4 642	4 728	4 816	4 906	4 998
Consumption	kt pw	3 850	3 999	4 147	4 253	4 343	4 427	4 520	4 606	4 695	4 785	4 877
Stock changes	kt pw	4	0	0	0	0	0	0	0	0	0	0
Price[3]	USD/t	3 647	2 599	2 527	2 617	2 688	2 850	2 929	3 026	3 127	3 216	3 305
Developed countries												
Production	kt pw	2 464	2 468	2 629	2 699	2 746	2 788	2 836	2 882	2 920	2 964	2 998
Consumption	kt pw	623	624	644	653	659	665	673	682	685	691	697
Developing countries												
Production	kt pw	1 547	1 653	1 640	1 675	1 719	1 761	1 806	1 846	1 896	1 943	2 000
Consumption	kt pw	3 228	3 375	3 504	3 600	3 684	3 762	3 847	3 924	4 010	4 094	4 179
OECD[2]												
Production	kt pw	2 696	2 719	2 877	2 949	2 998	3 044	3 096	3 144	3 187	3 233	3 272
Consumption	kt pw	899	912	940	956	969	984	999	1 015	1 026	1 039	1 053
Stock changes	kt pw	4	0	0	0	0	0	0	0	0	0	0
WHEY POWDER												
Wholesale price, United States[4]	USD/t	1 188	868	879	964	1 009	1 054	1 044	1 107	1 156	1 205	1 244
CASEIN												
Price[5]	USD/t	8 657	7 758	7 176	7 720	7 860	8 338	8 535	8 818	9 146	9 310	9 483

Note: Calendar year: Year ending 30 June for Australia and 31 May for New Zealand in OECD aggregate.
Average 2013-15est: Data for 2015 are estimated.

1. FOB export price, non-fat dry milk, 1.25% butterfat, Oceania.
2. Excludes Iceland but includes all EU28 member countries.
3. FOB export price, WMP 26% butterfat, Oceania.
4. FOB export price, sweet whey non-hygroscopic, Western Europe.
5. Export price, New Zealand.

Source: OECD/FAO (2016), "OECD-FAO Agricultural Outlook", *OECD Agriculture statistics* (database). doi: dx.doi.org/10.1787/agr-outl-data-en

StatLink http://dx.doi.org/10.1787/888933382199

Table 3.A1.7. **World fish and seafood projections**

Calendar year

		Average 2013-15est	2016	2017	2018	2019	2020	2021	2022	2023	2024	2025
FISH[1]												
World												
Production	kt	166 889	170 861	173 619	177 120	181 393	185 698	186 522	188 429	190 869	193 424	195 911
of which aquaculture	kt	73 305	77 708	80 214	83 531	87 527	91 583	94 404	95 257	96 941	99 425	101 768
Consumption	kt	166 187	170 782	173 546	177 043	181 322	185 623	186 444	188 356	190 792	193 344	195 827
of which for food	kt	146 648	151 503	154 286	157 788	162 071	166 357	168 542	170 098	172 328	175 040	177 679
of which for reduction	kt	15 623	15 359	15 426	15 498	15 580	15 680	14 402	14 834	15 116	15 031	14 951
Price												
Aquaculture[2]	USD/t	2 145.2	2 082.8	2 002.7	1 931.9	1 831.5	1 803.5	1 849.1	1 970.5	2 035.7	2 115.5	2 184.3
Capture[3]	USD/t	1 527.0	1 485.2	1 469.7	1 457.1	1 432.5	1 428.8	1 451.0	1 483.6	1 511.1	1 572.1	1 634.5
Product traded[4]	USD/t	2 866.7	2 722.4	2 624.0	2 540.0	2 412.0	2 370.0	2 428.0	2 539.0	2 595.0	2 661.0	2 719.0
Developed countries												
Production	kt	29 018	29 198	29 165	29 215	29 353	29 430	29 433	29 326	29 247	29 251	29 305
of which aquaculture	kt	4 393	4 591	4 677	4 808	5 021	5 227	5 332	5 305	5 319	5 412	5 521
Consumption	kt	36 748	36 542	36 468	36 499	36 784	36 981	37 077	36 996	37 294	37 608	38 045
of which for food	kt	31 917	31 678	31 778	31 880	32 231	32 500	32 635	32 673	33 065	33 441	33 950
of which for reduction	kt	4 387	4 432	4 270	4 211	4 156	4 096	4 070	3 953	3 861	3 801	3 732
Developing countries												
Production	kt	137 871	141 663	144 454	147 905	152 040	156 267	157 090	159 103	161 622	164 173	166 606
of which aquaculture	kt	68 911	73 117	75 537	78 723	82 507	86 356	89 073	89 953	91 622	94 013	96 247
Consumption	kt	129 439	134 240	137 079	140 544	144 539	148 642	149 367	151 360	153 499	155 735	157 781
of which for food	kt	114 732	119 825	122 509	125 908	129 839	133 857	135 907	137 425	139 264	141 599	143 730
of which for reduction	kt	11 235	10 927	11 156	11 286	11 424	11 584	10 332	10 881	11 255	11 230	11 220
OECD												
Production	kt	31 135	31 192	31 369	31 485	31 635	31 773	31 493	31 540	31 651	31 727	31 842
of which aquaculture	kt	6 165	6 457	6 549	6 677	6 932	7 196	7 358	7 344	7 376	7 500	7 628
Consumption	kt	38 680	38 996	39 153	39 258	39 587	39 838	39 809	39 870	40 306	40 710	41 227
of which for food	kt	32 314	32 600	32 777	32 958	33 370	33 686	33 855	33 940	34 411	34 850	35 410
of which for reduction	kt	5 827	5 833	5 824	5 758	5 684	5 629	5 442	5 417	5 381	5 348	5 304
FISHMEAL[5]												
World												
Production	kt	4 436.9	4 506.7	4 626.0	4 723.5	4 819.3	4 914.3	4 654.9	4 846.3	4 991.0	5 047.4	5 103.4
from whole fish	kt	3 164.9	3 166.2	3 207.9	3 241.8	3 274.9	3 310.9	3 002.9	3 125.4	3 208.2	3 198.1	3 188.8
Consumption	kt	4 523.4	4 527.1	4 534.7	4 664.5	4 770.5	4 893.7	4 945.2	4 730.3	4 906.0	4 982.5	5 077.1
Variation in stocks	kt	-86.5	-20.3	91.3	59.1	48.8	20.6	-290.3	116.0	85.0	64.8	26.3
Price[6]	USD/t	1 671.0	1 397.2	1 101.1	1 114.7	1 156.5	1 203.2	1 383.4	1 325.5	1 360.0	1 388.9	1 435.2
Developed countries												
Production	kt	1 351.7	1 483.1	1 484.9	1 507.6	1 535.0	1 557.7	1 586.3	1 596.9	1 612.8	1 636.0	1 657.9
from whole fish	kt	893.7	915.1	886.2	877.7	869.4	860.3	858.5	837.7	822.5	813.4	802.6
Consumption	kt	1 887.3	1 711.6	1 693.4	1 705.8	1 709.7	1 724.9	1 693.4	1 608.8	1 639.8	1 629.7	1 627.8
Variation in stocks	kt	-32.3	27.7	22.3	10.1	9.8	6.6	-79.3	41.0	20.0	14.8	8.3
Developing countries												
Production	kt	3 085.2	3 023.7	3 141.1	3 215.9	3 284.4	3 356.6	3 068.6	3 249.4	3 378.3	3 411.4	3 445.6
from whole fish	kt	2 271.2	2 251.1	2 321.7	2 364.0	2 405.5	2 450.6	2 144.4	2 287.7	2 385.6	2 384.7	2 386.2
Consumption	kt	2 636.1	2 815.4	2 841.4	2 958.6	3 060.8	3 168.8	3 251.8	3 121.4	3 266.3	3 352.8	3 449.4
Variation in stocks	kt	-54.2	-48.0	69.0	49.0	39.0	14.0	-211.0	75.0	65.0	50.0	18.0
OECD												
Production	kt	1 580.7	1 731.7	1 765.8	1 785.0	1 806.5	1 826.9	1 815.1	1 843.1	1 868.6	1 894.2	1 918.5
from whole fish	kt	1 134.6	1 180.8	1 186.6	1 177.1	1 165.4	1 156.5	1 116.8	1 115.9	1 112.9	1 108.6	1 102.8
Consumption	kt	2 038.7	1 836.1	1 822.7	1 844.3	1 856.0	1 880.8	1 854.4	1 771.4	1 812.4	1 808.2	1 812.2
Variation in stocks	kt	-41.2	42.7	47.3	15.1	4.8	6.6	-124.3	61.0	40.0	19.8	13.3

StatLink http://dx.doi.org/10.1787/888933382202

Table 3.A1.7. **World fish and seafood projections** *(cont.)*

Calendar year

		Average 2013-15est	2016	2017	2018	2019	2020	2021	2022	2023	2024	2025
FISH OIL[5]												
World												
Production	kt	857.8	918.1	933.8	948.9	965.5	981.8	935.0	966.1	990.0	997.7	1 005.6
from whole fish	kt	531.4	591.0	594.6	597.5	600.8	604.8	546.2	565.6	577.7	573.4	569.2
Consumption	kt	862.7	912.3	917.1	937.5	952.2	976.5	983.8	964.3	976.4	981.2	998.2
Variation in stocks	kt	-4.9	5.8	16.7	11.4	13.4	5.3	-48.8	1.8	13.6	16.6	7.4
Price[7]	USD/t	1 922.3	1 913.1	1 704.2	1 665.5	1 679.5	1 730.6	1 966.8	1 826.4	1 830.6	1 837.6	1 862.7
Developed countries												
Production	kt	390.1	399.4	400.4	406.2	413.7	419.6	426.6	429.4	433.2	438.8	444.1
from whole fish	kt	170.5	184.7	177.5	175.2	173.2	170.9	170.0	165.1	161.1	158.7	155.8
Consumption	kt	533.3	593.6	568.9	574.0	583.0	600.3	617.3	584.0	590.6	591.5	602.9
Variation in stocks	kt	-11.6	10.8	3.7	3.4	3.4	3.3	-15.8	0.8	2.6	2.6	2.4
Developing countries												
Production	kt	467.7	518.7	533.4	542.6	551.9	562.1	508.4	536.8	556.7	558.9	561.6
from whole fish	kt	360.9	406.3	417.1	422.3	427.6	433.9	376.1	400.5	416.5	414.7	413.4
Consumption	kt	329.3	318.7	348.2	363.5	369.1	376.2	366.5	380.3	385.8	389.7	395.3
Variation in stocks	kt	6.7	-5.0	13.0	8.0	10.0	2.0	-33.0	1.0	11.0	14.0	5.0
OECD												
Production	kt	489.4	504.5	514.0	520.8	528.5	536.0	537.3	545.7	553.4	561.7	569.7
from whole fish	kt	237.4	255.7	255.1	251.7	248.1	245.3	236.7	235.4	233.4	231.6	229.4
Consumption	kt	656.0	687.3	679.8	693.2	706.3	727.7	738.6	715.2	724.0	725.9	739.0
Variation in stocks	kt	-18.8	20.8	13.7	8.4	3.4	0.3	-22.8	0.8	3.6	6.6	2.4

Note: The term "fish" indicates fish, crustaceans, molluscs and other aquatic animals, but excludes aquatic mammals, crocodiles, caimans, alligators and aquatic plants.
Average 2013-15est: Data for 2015 are estimated.

1. Data are in live weight equivalent.
2. World unit value of aquaculture fisheries production (live weight basis).
3. FAO estimated value of world ex vessel value of capture fisheries production excluding for reduction.
4. World unit value of trade (sum of exports and imports).
5. Data are in product weight.
6. Fishmeal, 64-65% protein, Hamburg, Germany.
7. Fish oil, any origin, N.W. Europe.

Source: OECD/FAO (2016), "OECD-FAO Agricultural Outlook", *OECD Agriculture statistics* (database). doi: dx.doi.org/10.1787/agr-outl-data-en

StatLink http://dx.doi.org/10.1787/888933382202

Table 3.A1.8. **World biofuel projections**

		Average 2013-15est	2016	2017	2018	2019	2020	2021	2022	2023	2024	2025
ETHANOL												
World												
Production	mln L	111.5	119.3	122.0	123.2	124.2	125.1	125.1	125.7	126.4	128.0	128.4
of which maize based	mln L	59.2	62.9	64.9	64.8	64.4	64.2	63.3	63.1	63.0	63.4	62.6
of which sugar cane based	mln L	26.9	29.5	29.9	30.3	31.1	31.8	32.4	33.0	33.4	34.1	34.6
of which biomass based	mln L	0.4	0.6	0.6	0.6	0.6	0.6	0.6	0.6	0.5	0.5	0.5
Consumption	mln L	111.5	119.9	123.1	124.4	125.3	126.1	126.0	126.6	127.3	128.8	129.2
of which fuel use	mln L	89.0	96.5	99.4	100.3	100.8	101.2	100.7	101.0	101.2	102.4	102.3
Exports	mln L	7.3	7.7	7.8	8.0	7.8	8.2	8.0	7.8	7.7	7.4	6.9
Price[1]	USD/t	57.8	46.7	48.8	49.5	50.5	52.7	54.0	56.5	58.2	60.9	60.3
Developed countries												
Production	mln L	66.5	71.3	73.3	73.7	73.4	73.1	72.1	71.6	71.4	71.8	71.0
Consumption	mln L	68.0	71.9	74.3	74.8	74.6	74.4	73.2	72.7	72.3	72.9	72.2
of which fuel use	mln L	61.6	65.4	67.7	68.3	68.1	67.9	66.7	66.2	65.8	66.4	65.6
Net trade	mln L	-1.4	-0.4	-0.9	-1.1	-1.2	-1.3	-1.1	-1.0	-0.9	-1.1	-1.2
Developing countries												
Production	mln L	45.1	48.0	48.6	49.5	50.8	52.0	53.0	54.1	55.0	56.2	57.4
Consumption	mln L	43.5	48.0	48.9	49.6	50.7	51.7	52.8	53.9	54.9	55.9	57.0
of which fuel use	mln L	33.7	37.9	38.7	39.3	40.4	41.2	42.2	43.3	44.2	45.1	46.2
Net trade	mln L	1.0	-0.2	0.4	0.6	0.7	0.7	0.5	0.5	0.4	0.6	0.6
OECD[2]												
Production	mln L	64.8	69.6	71.6	71.9	71.6	71.3	70.3	69.8	69.6	69.9	69.1
Consumption	mln L	66.8	70.8	73.1	73.6	73.4	73.1	71.9	71.4	71.1	71.6	70.9
of which fuel use	mln L	60.8	64.7	67.0	67.5	67.3	67.0	65.8	65.3	64.9	65.5	64.7
Net trade	mln L	-1.9	-1.0	-1.5	-1.7	-1.8	-1.8	-1.7	-1.6	-1.5	-1.7	-1.7
BIODIESEL												
World												
Production	mln L	31.1	33.2	34.5	35.3	36.7	37.9	38.8	39.6	40.2	40.8	41.4
of which vegetable oil based	mln L	25.2	26.3	26.6	26.9	27.5	28.4	29.0	29.3	29.5	29.8	30.1
of which waste based	mln L	2.4	2.9	3.4	3.7	4.2	4.4	4.7	5.1	5.4	5.8	6.0
of which biomass based	mln L	0.0	0.0	0.0	0.0	0.0	0.0	0.0	0.0	0.0	0.0	0.0
Consumption	mln L	30.3	33.5	34.7	35.5	36.9	38.1	39.0	39.8	40.4	41.0	41.6
Exports	mln L	4.0	2.2	2.6	2.7	2.7	2.7	2.7	2.7	2.7	2.7	2.6
Price[3]	USD/t	93.9	72.1	71.9	73.7	76.8	81.5	85.9	87.3	87.1	88.4	88.4
Developed countries												
Production	mln L	18.1	19.4	19.9	20.3	21.1	21.7	22.0	22.3	22.4	22.4	22.4
Consumption	mln L	19.9	20.7	21.5	22.0	22.9	23.4	23.7	24.0	24.0	24.1	24.1
Net trade	mln L	-1.8	-1.3	-1.6	-1.7	-1.7	-1.7	-1.7	-1.7	-1.7	-1.7	-1.7
Developing countries												
Production	mln L	13.0	13.9	14.6	15.0	15.5	16.3	16.8	17.3	17.8	18.4	18.9
Consumption	mln L	10.4	12.7	13.2	13.5	14.0	14.7	15.3	15.8	16.4	16.9	17.5
Net trade	mln L	2.6	1.1	1.4	1.5	1.5	1.5	1.5	1.5	1.5	1.5	1.4
OECD[2]												
Production	mln L	18.7	20.1	20.7	21.1	22.0	22.5	22.9	23.1	23.2	23.3	23.3
Consumption	mln L	20.3	21.3	22.2	22.7	23.5	24.1	24.5	24.7	24.8	24.8	24.8
Net trade	mln L	-1.7	-1.2	-1.5	-1.6	-1.6	-1.6	-1.6	-1.6	-1.5	-1.5	-1.5

Note: Average 2013-15est: Data for 2015 are estimated.

1. Wholesale price, United states, Omaha.
2. Excludes Iceland but includes all EU28 member countries.
3. Producer price Germany net of biodiesel tariff and energy tax.

Source: OECD/FAO (2016), "OECD-FAO Agricultural Outlook", *OECD Agriculture statistics* (database). doi: dx.doi.org/10.1787/agr-outl-data-en

StatLink *http://dx.doi.org/10.1787/888933382219*

Table 3.A1.9. **World cotton projections**

Marketing year

		Average 2013-15est	2016	2017	2018	2019	2020	2021	2022	2023	2024	2025
WORLD												
Production	Mt	24.9	23.6	23.9	24.3	24.7	25.2	25.8	26.4	27.1	27.6	28.0
Area	Mha	32.7	31.2	31.5	31.6	31.7	32.0	32.3	32.5	32.8	32.9	33.1
Yield	t/ha	0.72	0.76	0.76	0.77	0.78	0.79	0.80	0.81	0.83	0.84	0.85
Consumption[1]	Mt	23.9	24.7	25.2	25.5	25.9	26.2	26.6	26.9	27.4	27.8	28.3
Exports	Mt	8.1	7.3	7.5	7.6	7.7	7.8	7.9	8.1	8.3	8.5	8.7
Closing stocks	Mt	20.2	18.8	17.5	16.3	15.1	14.1	13.3	12.8	12.5	12.2	11.8
Price[2]	USD/t	1 699.1	1 280.3	1 239.7	1 300.6	1 344.8	1 432.4	1 486.4	1 528.1	1 524.0	1 501.6	1 479.4
DEVELOPED COUNTRIES												
Production	Mt	5.4	5.3	5.3	5.5	5.5	5.6	5.8	6.0	6.1	6.3	6.4
Consumption	Mt	1.7	1.8	1.8	1.9	1.9	1.9	1.9	2.0	2.0	2.0	2.1
Exports	Mt	4.2	3.7	3.9	4.0	4.0	4.1	4.2	4.3	4.5	4.6	4.7
Imports	Mt	0.4	0.4	0.4	0.4	0.4	0.4	0.4	0.4	0.4	0.4	0.4
Closing stocks	Mt	1.5	1.5	1.5	1.6	1.5	1.6	1.6	1.6	1.6	1.6	1.6
DEVELOPING COUNTRIES												
Production	Mt	19.5	18.3	18.6	18.8	19.1	19.6	20.0	20.5	20.9	21.3	21.6
Consumption	Mt	22.2	22.9	23.3	23.6	24.0	24.3	24.6	25.0	25.4	25.8	26.2
Exports	Mt	3.9	3.6	3.6	3.6	3.6	3.7	3.7	3.8	3.9	3.9	4.0
Imports	Mt	7.6	7.0	7.1	7.2	7.3	7.4	7.6	7.8	8.0	8.1	8.3
Closing stocks	Mt	18.7	17.3	16.0	14.7	13.5	12.5	11.7	11.2	10.8	10.5	10.2
OECD[3]												
Production	Mt	4.9	4.7	4.8	4.9	5.0	5.0	5.2	5.3	5.5	5.7	5.8
Consumption	Mt	3.3	3.4	3.4	3.4	3.4	3.4	3.5	3.5	3.5	3.5	3.6
Exports	Mt	3.3	2.9	3.0	3.1	3.2	3.2	3.4	3.5	3.6	3.8	3.9
Imports	Mt	1.6	1.7	1.7	1.7	1.7	1.7	1.7	1.7	1.7	1.7	1.7
Closing stocks	Mt	1.7	1.7	1.7	1.7	1.7	1.7	1.7	1.7	1.7	1.7	1.7

Note: Marketing year: See Glossary of Terms for definitions.
Average 2013-15est: Data for 2015 are estimated.

1. Consumption for cotton means mill consumption and not final consumer demand.
2. Cotlook A index, Middling 1 3/32", c.f.r. far Eastern ports (August/July).
3. Excludes Iceland but includes all EU28 member countries.

Source: OECD/FAO (2016), "OECD-FAO Agricultural Outlook", *OECD Agriculture statistics* (database). doi: dx.doi.org/10.1787/agr-outl-data-en

StatLink http://dx.doi.org/10.1787/888933382229

OECD PUBLISHING, 2, rue André-Pascal, 75775 PARIS CEDEX 16
(51 2016 02 1 P) ISBN 978-92-64-25322-3 – 2016

www.ingramcontent.com/pod-product-compliance
Ingram Content Group UK Ltd.
Pitfield, Milton Keynes, MK11 3LW, UK
UKHW051505290926
15471UKWH00008B/1438